Emergence, Complexity and Computation

Volume 11

For further volumes:
http://www.springer.com/series/10624

About this Series

The Emergence, Complexity and Computation (ECC) series publishes new developments, advancements and selected topics in the fields of complexity, computation and emergence. The series focuses on all aspects of reality-based computation approaches from an interdisciplinary point of view especially from applied sciences, biology, physics, or Chemistry. It presents new ideas and interdisciplinary insight on the mutual intersection of subareas of computation, complexity and emergence and its impact and limits to any computing based on physical limits (thermodynamic and quantum limits, Bremermann's limit, Seth Lloyd limits…) as well as algorithmic limits (Gödel's proof and its impact on calculation, algorithmic complexity, the Chaitin's Omega number and Kolmogorov complexity, non-traditional calculations like Turing machine process and its consequences,…) and limitations arising in artificial intelligence field. The topics are (but not limited to) membrane computing, DNA computing, immune computing, quantum computing, swarm computing, analogic computing, chaos computing and computing on the edge of chaos, computational aspects of dynamics of complex systems (systems with self-organization, multiagent systems, cellular automata, artificial life,…), emergence of complex systems and its computational aspects, and agent based computation. The main aim of this series it to discuss the above mentioned topics from an interdisciplinary point of view and present new ideas coming from mutual intersection of classical as well as modern methods of computation. Within the scope of the series are monographs, lecture notes, selected contributions from specialized conferences and workshops, special contribution from international experts.

Ali Sanayei · Otto E. Rössler

Chaotic Harmony

A Dialog about Physics, Complexity and Life

Springer

Ali Sanayei
Institute for Theoretical Physics
University of Tübingen
Tübingen
Germany

Otto E. Rössler
Institute of Physical and Theoretical Chemistry
University of Tübingen
Tübingen
Germany

ISSN 2194-7287 ISSN 2194-7295 (electronic)
ISBN 978-3-319-06780-3 ISBN 978-3-319-06781-0 (eBook)
DOI 10.1007/978-3-319-06781-0
Springer Cham Heidelberg New York Dordrecht London

Library of Congress Control Number: 2014940404

Printed on acid-free paper

Springer is part of Springer Science+Business Media (www.springer.com)

Foreword

Ali and Otto first met in 2011 at a conference in Greece on dynamical systems and chaos. At this meeting, the two talked at length mostly on chaos theory, and from then on they remained in regular contact discussing different topics. In December 2012, Ali, a young physicist, decided to capture different topics and ideas in Otto's research activities of more than 40 years in a biographical interview. So, on weekends Ali brought a video camera for the interview. Even though the red thread of the interview was intended to be autobiographical, the two of them rather turned to eagerly discuss unsolved problems in science as a whole. While for Ali quantum mechanics and classical mechanics are two powerful streams in modern physics—he always tries to find an analogy or proportionality between them–, in Otto's view both fields do not need to be seen as fundamentally separate whereby he draws a connection to the ancient Greek philosophers. Something the two of them have in common, however, are a high opinion of Paul Dirac's work, as well as of Graham Farmelo's colourful biography.

In the hour-long interviews of Ali and Otto, life and biography somehow are eclipsed by the debate about unsolved problems and complex theoretical themes. Both discussion partners try to find explanations, whereby Ali defends quantum mechanics, which Otto attempts to incorporate into his web of classical assumptions and interpretations. While at first only a few interviews had been planned, the two scientists came together for more and more sessions, further illustrating their love for finding answers to physical problems. Otto is inclined to mix in different disciplines because of his long research experience in science, with some cherished topics repeated like musical themes. Evolution is one of these beloved topics and chaos theory, of course, is another. In addition, Otto tries to sell thoughts about endophysics and cryodynamics, newly designed theories or disciplines, as he dares to call them. In Otto's eyes, these new fields have major implications for the understanding of physics and cosmology, maybe even harbouring new sources of energy.

The big topics focused on in 25 sessions are chemical evolution (life), humanity theory, brain equation, chaos, endophysics, and cosmology. In Otto's opinion, many points are already solved, but of course, as unassuming as he is, he knows that there still is much work to be done and that many more convincing arguments are necessary to win over the physics community. Otto is very thankful for his many co-operations and long-term associations with lifelong friends across the globe and also with his many students. The gratefulness for the privilege to do research and teach at the university is evident. His main tools are paper and pencil and in our time, of course, the computer, although "only" to demonstrate his ideas. Ali is very much engaged in the roots of physics and natural science as a whole. Perhaps he sometimes is too tolerant and too hesitant to interrupt Otto's repetitions and excitement in explaining his beloved topics. On the other hand, Ali is very determined to get the right point and to discover surprises in reflecting new topics of science. Both, the young and the almost old scientist in dialogue, reveal an optimistic view, playful intuition, enthusiasm and joy in their joint—sometimes a little chaotic—journey through the natural sciences. And so, in the words of Heraclitus "out of differences the most beautiful harmony arises."

Reimara Rössler
Ivan Zelinka

Preface

Scientific dialogs in a genuine atmosphere surrounded by creative paintings above a piano, with beautiful various *objets d'art*, fantastic photographs and many prestigious books bring to life the eminent Rumi's statement that "There is no limit for thought, and there is a world of freedom inside every person." That was the atmosphere in which I conducted 25 interviews with a scientist who ignited his own "Otto-cycle" in the 1950s, to become a biologist at first, and turned to theoretical chemistry afterwards. Later, Maxwell's equations electrified him, Poincaré's three-body problem and chaotic strange attractors absorbed him, and Einstein's general theory of relativity attracted him to physics. Nevertheless, at the end of my interviews as long as the "cycle" was running with its high efficiency, he could also become a theologian and ultimately even combine all of those four strands together!

The pivotal principle for Otto E. Rössler and me, which made both of us have 25 prosperous dialogs, was Dirac's Principle of Beauty: "If one is working from the point of view of getting beauty in one's equations, and if one has really a sound insight, one is on a sure line of progress." Let it not remain unsaid that the elegant atmosphere in which those talks took place supported metaphysically our joint principle and even led us spontaneously to talk about a link between symmetry and art.

It was not the first time that I had discussions with Otto, but it was the first time that we had dialogs in that exceptional atmosphere. Although our discussions were not "dialogs concerning the two chief world systems," we did not achieve complete agreement on several ideas. Nonetheless, for those special interviews I had made a treaty with myself, that I should not criticize him during his elaborations. Although it was quite difficult on several occasions to be faithful to my inner treaty, I succeeded in not having a pro and con round-table with him but rather an interview. The main reason for my promise was simply to get to know his complete ideas about a given subject without any external intervention. Otherwise due to his kindness, as soon as I wanted to make a comment he would stop talking in a flow and might lose a main clue and then it became difficult to come back to the primary path. In this way, I can now say that I have obtained almost complete answers to my questions, none of which he knew beforehand. One could imagine that it may be burdensome to undergo many interviews and to answer various questions for which one does not have any idea in advance. So the next point which I should perhaps underline here is that those dialogs could not only collect comparatively complete answers, but also that Otto did not have any idea what my questions would be. It is also worth mentioning that I personally had no idea what his answers would be so that I could have tailored my questions in anticipation. Such a strategy—despite its seeming laboriousness—could enlarge the degree of creativity in our project and helped both of us not to saturate in a steady state.

The main thread in these dialogs was to start out from an early age, when Otto E. Rössler began becoming a scientist, up until now. Accordingly, the style of the book can be called a "scientific autobiography," for we were not going to talk about the private life. As I see it, one of the principal advantages of such a collection is to harvest his ideas about several subjects which can stimulate and foster more discussions and trigger comments. So I would be delighted if I had succeeded in establishing a basis for others who would like to discuss with Otto or even try to refute his ideas. As physicist John Wheeler, and subsequently also philosopher of science, Karl Popper, put it, "our whole problem is to make the mistakes as fast as possible"; we also are reminded of Einstein's saying that the essence of science is that "anyone who has never made a mistake has never tried anything new."

To preserve the dialogical structure, we kept the style of the whole book in the form of "questions and answers." Moreover, the reader will see an easy switching of subjects between biology, theoretical chemistry and physics. Keeping the autobiographical structure of the book implied to preserve the sequence of themes as they occurred. Otto loved to remain in his "cycle" and draw upon various fields and combine them in unexpected ways to answer my questions. Hence the reader will realize that when Otto talks about, for example, intelligence, he cannot reign-in his emotions and starts explaining the special behavior of orangutans with abundant eagerness! Due to his special style, he also mentions many names and works—not only in physics, but also in biology, zoology, chemistry, philosophy, mathematics, literature, theology, etc. In order that the reader can easily follow, a comprehensive index and bibliography have been compiled at the end.

Johann Wolfgang von Goethe believed that "Every day we should hear at least one little song, read one good poem, see one exquisite picture, and if possible, speak a few sensible words." I hope our 25 dialogs can be considered as some "sensible words."

My special gratitude goes to Oliver Purnell and Michal Rössler for editing all interviews. I would also very much like to thank Ivan Zelinka, Guanrong (Ron)

Chen, Thomas Ditzinger and Andrew Adamatzky who helped me a lot to finish this work. I am delighted to dedicate the book to my parents who first taught me that, in any chaos, one can eventually find harmony.

Tübingen, December 2013 Ali Sanayai

Contents

Chapter 1
Early Steps in Science

My first question: When did you first encounter the philosophy of science? When was that important point in your life?

I was a playful child, and as a child I had a "Märklin Baukasten," a metal construction kit for children to build mechanical toys with wheels, and I invented a little motor using a rubber band so that the rubber band would be coiled up on the axis of a simple toy car. The car would run a little while on its own, it was absolutely trivial, but I loved these things. Later, out of this familiarity with metal plates and axes and cogwheels I came up with a strange idea which I did not understand at the time. If you had a vertical metal plate and through that plate there came out 2 axes with a turning wheel each as a knob, and there was a hidden mechanism connecting the wheels with each other, I convinced myself that from the outside, you could not judge for sure what kind of mechanism was hidden behind these two wheels. At that time I did not know anything about measurement or uncertainty or Gödel, but I was fascinated by the kind of uncertainty that would arise—this just came back to me. I am not sure that my father understood what I had in mind when I confronted him with this problem at the time.

Do you remember the exact time or year?

This was when I was 11 years old, in 1951. Shortly thereafter, this equipment for building-up mechanical toys was complemented by electrical components, including little electromagnets, so that a DC motor could be built in the same hands-on fashion out of Märklin elements. There was also a little book edited by a firm to encourage young people to work with the toy elements they provided. I remember I wrote a letter to this firm named "Rim" after correcting all the misprints and errors I had found in the booklet they had supplied with the material. I got a very kind letter back from them thanking me for having helped them improve their book. I did not know at that time that this was typical scientific behavior, displayed. It was just my childish joy in solving mechanical problems and inventing a machine. Then I had this catastrophe in life called puberty. Afterwards, radio-amateuring became my favorite pastime. I became the youngest licensed radio amateur in the country. I had invented by trial and error a new

A. Sanayei and O. E. Rössler, *Chaotic Harmony*,
Emergence, Complexity and Computation 11, DOI: 10.1007/978-3-319-06781-0_1,

amplitude modulation for radio transmitters whereby the amplitude of the carrier wave was made proportional to the sound wave, a fact which would cause a strong distortion as one would guess. But I included a negative feedback—from the demodulated carrier wave back to the low-frequency input—to effectively suppress the distortion despite the fact that the carrier wave still mirrored the voice volume. In this way, one could save a lot of money in terms of the amount of electricity consumed by a transmitter. This was my first invention if you so wish.

The year was?

That was when I was 17 years old.

When working with this equipment, were you also dealing with philosophical questions?

This had just been "stupid science" as it were,—mechanical and electrical engineering. Then there occurred something like a cut in my life, I discovered the poet Schiller in high school. I saw that he was interested in deep questions regarding the motivations of human beings. So I started trying to understand how motivations work, and the question of free will appeared very important to me. Is not everything deterministic, but how then can you have a free will? When in high school still, I was asked to write a long essay on a favorite topic of mine over the summer holidays, I wrote about Schiller the poet and the problem of free will in his plays. I saw that he had a very deep understanding of human motivations and that one could derive logically that the "net of motivations" was infinitely complex. I wonder how I came to this idea at the time. My high school teacher signed the essay but refused to grade it. The text contained also religious ideas. I drew a connection between Schiller, religion and philosophy. At that time I had entered a religious order on my own, the so-called 3rd order of Saint Francis. I was under the influence of an impressive Fransciscan monk who was very short, had a caring personality and who, when preaching, stood like a giant. Later I would ask him whether he would approve of my becoming a scientist or whether I should join the first order as he had hoped. He told me that it was okay if I continued with my deep insights. I did not know yet that Teilhard de Chardin, a Jesuit priest, had hit earlier on the very question which I felt I had solved.

What exactly was that question?

The question was, why do we have life and why do we exist? Is there something lawful behind this? Can this be fully understood? I was optimistic that one could completely understand everything, including the origin of life—a mechanism which predicts why life must arise and what is the "drive" behind it. What the force is that is giving rise to life.

Do you know the exact year when you solved this question?

Succeeding completely to my own satisfaction came later, when I was 20 years old, in the third semester of my medical studies.

And to answer this question: You didn't try to study theology, but went through science instead?

I attended theological courses at the university, and it was a theological motivation that spurred me on, but the mechanism envisioned was purely scientific. I effectively re-discovered Teilhard de Chardin's finding: the existence of a drive that pervades the universe and leads towards "Point Omega," the end point of evolution. At the same time, I liked to have a mechanism for this energetic drive which involved a different type of thinking. I saw both ways in one glance and was sure they would meet. Specifically, there was an insight I got from a little textbook that every medical student got for free from a pharmaceutical firm at the time called Tropon. It was about physiology and was written by Hans Lullies. It featured a person whom I would meet later in life, Ludwig von Bertalanffy. He effectively opened my eyes to a new principle. Bertalanffy was a theoretical biologist and had written the first book titled "Theoretical Biology," as I later learned. He also had invented the notion of a "Fliessgleichgewicht"—literally "flow-based equilibrium"—which is misleadingly translated into "steady state." He was quoted in the little book with a mental picture offered, a juxtaposition of fire and life. A mouse would, if put into a "black box" with only input and output measured, be virtually indistinguishable from a candle flame. This inspired me to see a connection between the flame as a self-maintaing dynamical structure and life. It made me see the unbounded power of the flame—to transform everything it touches into itself. I saw the connection between flame and life, and I also could see the difference: The flame cannot become something else, it can only become more of the same, more fire. The juxtaposition allowed me to see something in between the two: rust. Rust causing iron to become rusty is also an autocatalytic process; this is what rust and flame have in common. However, from experience one knows that rust looks different here and there. While being autocatalytic and trying to become more and more of the same thing, rust at the same time has more degrees of freedom. So, I could see that if you combine autocatalysis with more chemical degrees of freedom, then maybe this built-in power of self-amplifying autocatalysis in chemistry (if given free reign) makes for a bridge between the primitive form of autocatalysis seen in fire and the highly developed branching autocatalysis that we find in life. I predicted that this drive that I had glimpsed in a Teilhard-type fashion would be inherent in chemistry, so that a very simple chemical soup would self-organize in such a way that eventually, new side-paths of the autocatalysis would be opened-up that are themselves autocatalytic, but more sophisticated. Charles Darwin had had essentially the same idea 100 years before. So it was not very original, but the "pictorial conviction" that this really carries through was perhaps new.

You had this idea when you were 20 years old and a medical student?

Yes, I got convinced I could prove that life was a self-organizing process.

But the motivation came from theology. Why did you go to medicine, why not theology or, for example, physics?

It was two reasons, I think. When building receivers and transmitters on my own recipes, I had started to read electronics engineering books. Once, I encountered a book with equations meant to describe how these elements, when put together, worked. On the left-hand side of the equation there was a dot on the variable x. I could not understand this equation at all and knew that I would never understand it. This was the first differential equation I ever encountered, and it would take me 10 years before I returned to differential equations. However, the reason I chose medicine was an ethical one. I thought, one has to devote one's life to serving other people, it was a kind of substitute for my not doing theology, for not becoming a begging monk. Something like practical theology. Then, while I studied medicine and physiology, this window opened-up to understanding evolution, I was kind of carried away from this philanthropic motivation into a causal scientific type of thinking which nonetheless enabled me to keep my benevolent intentions. I did not think there was a contradiction between these two ways of thinking and acting. But eventually, the scientific side would win the upper hand in my life, a fact which I would have been unable to predict at the time.

You began your studies of medicine at age 19. Was there an additional reason for choosing medicine? Did you get any suggestions?

My mother was very much trying to push me in this direction because her father had been a medical doctor after first studying theology.

You started at the University of Tübingen. How many years did it take to finish medicine? As a medical student, what was the next interesting question to work on?

It was a slow development. I turned to one of the best professors of medicine in Tübingen, Erich Letterer, head of the pathology institute. I became a doctoral student in pathology because I liked his style of lecturing. I worked in his institute for several years as a student, assisting him even to dissect colleagues at the university in which case he felt responsible to make the final diagnosis himself. I was assigned an immunological dissertation, trying to reproduce a certain type of disease in mice called Alzheimer's, by injecting them daily with a foreign protein, bovine gamma globulin, in differing doses. This took 4 years, a very long time. The mice did not produce the symptoms that were expected, but instead produced immunological tolerance, a new phenomenon I did not know about when describing it in my dissertation, but which would soon after win a Nobel prize for its first discoverers. It was a big disappointment to my professor that I had not found the disease he had expected—amyloidosis. The experience helped me overcome disappointment in science. While still working in the pathological institute at night, I started to write scientific letters. My motivation was that in this way I might be able to help a sick child I had met in the surgical clinic once. I was then a student in a higher semester at 24 years of age. The first letter, addressed to Ludwig von Bertalanffy, apparently never reached him—I met him 5 years later—, described my theory of evolution. Another letter went to physicist Carl-Friedrich von Weizsäcker who had written a book "The History of Nature" which contained similar ideas to those I had developed. A third went to Konrad Lorenz who had written nice books on animal behavior. The letter contained a proof that if adding consciousness to animal behavior had any evolutionary cost to it, consciousness would have been selected-out. I eventually received an answer almost a year later when I was already in Seewiesen. However, as stated I had initially written to Carl-Friedrich von Weizsäcker.

So you encountered physics during your medical studies?

Yes, I had read a book by von Weizsäcker. In one chapter of his pocket book "The History of Nature" he had commented on the origin of life. I could specify what he had said there. So I came into contact with this eminent physicist. Weizsäcker got interested and made the "mistake" of answering my letter. He invited me to visit him at the home of his mother near Lake Constance. It was a pleasant encounter, his charming mother brewed coffee. He told me things which I had not expected talking about, like that Lise Meitner had asked him whether she should publish, and he had encouraged her to do so. What I did not know at the time was that he had been the scientific assistant of Lise Meitner 's at her institute in Berlin, and that she had had to leave the country. Only later did I realize the historical context. Meitner did publish, but her colleague Otto Hahn had done so a month before. So everybody thinks he was the first, even though her article was submitted earlier. So I suddenly was included a bit in this strange tragic history of physics, owing to this encounter and friendship with von Weizsäcker. He was a theoretical physicist and close friend of Werner Heisenberg's . He told me how Heisenberg had told him in a taxi cab—when Weizsäcker was only 13—that "I (Heisenberg) have just proven Newton wrong." This information that it is possible to prove a giant wrong triggered his desire to do science. They enjoyed a lifelong friendship with professional benefits, as Heisenberg would always make sure that Weizsäcker got a new position whenever he chose to enter a new field.

Weizsäcker also told me the story of the day Heisenberg visited Niels Bohr during the war in Copenhagen; he had accompanied Heisenberg waiting outside, while Heisenberg, talking to Niels Bohr, handed over the blueprint of a nuclear reactor. This fact does not seem to be very well known; at least Michael Frayn does not mention it in his play "Copenhagen." During wartime it is high treason if a secret is given to the other side and if I remember correctly, I learned that Heisenberg wore the uniform of an officer of the German army. I later asked a person who should know because of his close relationship with Heisenberg, and was told that this was not implausible because there was a law in Germany called "Sippenhaft", where the whole family would be put in jail if one member had done something against the state. But if Heisenberg had committed the same act as a member of the army, the army would have been responsible if the truth came out. So the story may be true. This is just politics, I apologize, but it shows how a young person can suddenly be told secrets of a historical dimension, and it illustrates my saying that science is friendship.

After that meeting, you went back to Tübingen and continued your medical studies?

I had told Weizsäcker that I wanted to pursue science from now on and stop studying medicine. But he replied, "Please, do me the favor and do your medical exam. You just need an academic degree, then you can continue doing science. If you do me that favor, I will give you a position at Max Planck." Indeed, after finishing I got a visiting position at Max Planck in Seewiesen near Starnberg.

When did you finish your dissertation?

I was one year late because the doctoral thesis took so long. I was 26 at the time. I found myself at Seewiesen giving a talk on the evolution of life, and no one was interested at the Cybernetics Institute. But there was a neighboring Max Planck Institute headed by Konrad Lorenz. While I was already working at the division of Mittelstaedt, I received a letter from Lorenz in the neighboring institute, sent to my student address in Tübingen, in reply to my letter from almost a year before. I could answer that I was working 500 yards from his place, so he had to invite me to pay him a visit out of curiosity or perhaps because my letter had caught his interest. As mentioned, I had given a proof that it is never the subjective side which is selected by evolution, only the functional side. So if one camel is more thirsty and the other has the better nose, it is the better nose that is picked by evolution, not the stronger thirst. That was a new thesis to Lorenz apparently.

Which Lorenz?

Right, it was ethologist Konrad Lorenz, not my later chaos friend meteorologist Ed Lorenz. I then went to see Konrad and that became a friendship, too.

What was the main topic in your thesis?

The title of my thesis on immunology, in fact immunological tolerance, was "Long-term immunization of albino mice with bovine gamma globulin."

When you worked at Max Planck—did you forget everything in medicine?

I was suddenly in the role of a theoretician of evolution.

So at Max Planck you did not work on physics, only on evolution?

Yes, on evolution as a physical phenomenon. I discussed mainly with Lorenz. He invited me to give a talk in his department about my understanding of how evolution works in nature. I gave the example of a glass filled with honey that is tipped-over on a table, and how the first outpouring wave causes a mound which inhibits the rest from flowing out freely. It was my example of how free energy could, for a while, remain hidden in pockets from which it would exert a further influence before being dissipated away, thereby improving the chances for more free energy to be drawn into round-about ways of dissipation, and so forth. Lorenz was quite taken with this presentation, other attendees from his institute were more critical. One advanced colleague, Norbert Bischof, realized that I did not know anything about differential equations and was the first to kindly suggest to me the use of a computer. He even made a brief attempt at teaching me programming—in vain.

You did not have any course on mathematics in medicine?

No, I only trusted my own intuition. These two famous people, von Weizsäcker and Lorenz, each made the mistake of taking seriously the intuition of someone who knew nothing in a formal sense. This type of credit given—call it friendship—is the essence of science, I still feel. I loved the questions found, no matter what the available tools, and became acquainted with the fundamentals in a stepwise manner.

You learned differential equations at that Institute?

Not at all, that took its time. I lost my position at Max Planck after 10 months because I did not fit squarely into the department of Biocybernetics. Strangely, my mother paid a visit and asked to extend my stay but she was only told "Your son will make his way."

Now you lost your position at Max Planck and your age was 26, so we'll stop here and will continue starting the next session at 26 years!

Chapter 2
Entering Theoretical Biology

You are now 26 years old. Our last point was that you lost your position at Max Planck because your stipend had finished. Now, please, continue: What happened after that?

I turned 27 that year. It was the year I would get married. I worked at the University of Marburg for a while to finish my medical training as a "Medizinalassistent." Before getting the approbation (license to work as a doctor) one had to work for 2 years as a supervised medical intern in different departments. In my spare time I continued with my scientific notes.

Is there anything important about getting married? From the Copenhagen viewpoint it is a spike in your psi-function.

I agree—many things happen when you make such decisions in your life, it is not the responsibility of one's own self alone, so the situation is very complex. You are at the mercy of another person in a symmetric, compounded way. It is a miracle. It is much easier to continue on science. After leaving Seewiesen I continued scribbling down notes on my thoughts on the origin of life. A short unpublished paper emerged. I also was allowed to give a talk on my theoretical approach to evolution in the medical clinic in Marburg in the lecture hall. It was received with interest. It was my first public talk after the two seminar talks held at Seewiesen. I learned to talk slowly. The auditorium felt that I seemed to belong to the 19th century—an opinion my wife heard. The origin of life remained my topic while working in medicine and I had also another meeting with Carl-Friedrich von Weizsäcker in Marburg. His encouragement continued.

What exactly was the project you were thinking about?

It had to do with Teilhard de Chardin. I realized that he had tried to build a consistent theory of evolution, based on a pervasive drive in nature seen by him that propels everything towards becoming more and more highly organized. He coined the notion of the "Point Omega"—the first attractor in history—which attracts everything in the cosmos (first spotted by him in 1915, when he was a chaplain at the front, under the name "Point Theta" at first—a bull's eye).

A. Sanayei and O. E. Rössler, *Chaotic Harmony*,
Emergence, Complexity and Computation 11, DOI: 10.1007/978-3-319-06781-0_2,

Everything in the universe is propelled towards an asymptotic state of a maximally highly organized complexity. I could support this idea of Teilhard's eventually with my chemical-kinetic equations.

Then I was allowed to visit the institute of a famous man, a freshly decorated Nobel Prize winner Manfred Eigen, who kindly invited me to give a talk—an event which was consummated before his whole institute on a skiing vacation in Hoch-Sölden in Austria. It was an important experience because I completely flunked the test. I gave a talk about my ideas on how chemical substances influence each other so that eventually you get autocatalysis in a network that is branching out. In the discussion I was asked whether I could define mathematically what a differential equation is. I answered that I had just presented many examples of such equations, what was it that was meant by defining it exactly? I was told that since I would not answer formally, it was no use talking to me any longer. Such an experience can make you stronger.

Who told you so?

The highly decorated scientist thereby effectively forced me to be more formal than I had been. Much later both he and I would independently write a paper on mad-cow disease. So our thoughts apparently went along parallel lines over decades.

And then you modeled your idea as an equation?

Yes, because a good fate intervened. I was using libraries very often and wrote a big paper on chemical evolution that by happenstance never appeared. I submitted it to the Journal of Theoretical Biology in Buffalo, N.Y. and got a favorable review but was told I should translate it from German into English even though the journal in principle accepted German manuscripts. My English was not good enough, I felt, a drawback I could not overcome at that time. Strangely, the same institution in Buffalo had posted an announcement for a stipend in the Journal of Theoretical Biology. Since I knew that my wife was interested in going to the U.S., I applied for this stipend. Later I learned that I was the only one out of about 170 applicants who got it. So by this stroke from heaven I suddenly found myself working at the Center for Theoretical Biology. It may have had to do with my paper submitted to the same institution. It gave me this opportunity to meet a whole new continent; America.

You were 27 years old? And how long did you stay at the institute in Buffalo?

It was only a bit more than half a year. It was in 1969, so I was 29 years old.

Did you do any special work during the 6 months there?

Actually no. My wife had a position at the famous Immunological Institute at the same university, with Professor Ernest Witebsky. I was just sitting in my room in the institute at Ridge Lea Road, writing notes and talking with people at the institute, mostly with Bob Rosen who had just finished the manuscript for his book

"Dynamical System Theory in Biology." I was reading these notes and we talked, and I was attending talks of visitors, including Ludwig von Bertalanffy who was there, too—to whom I had written my first scientific letter 5 years before which, as you may remember, he apparently never got. It was very nice to talk with him about evolution, the origin of life, theoretical biology and the way he had invented it as a science. I also attended a talk by Stu Kauffman at the Buffalo Institute. Being a medical doctor, too, he had independently invented almost the same theory of chemical evolution as I had. A lifelong friendship with Bob Rosen followed, and with Stu.

And then you came back from the United States to Germany?

Yes, after having been given a "semi-permanent position" in Buffalo which was a very nice thing to hear. I then got a stipend from the German DFG research organization. I chose to work in Tübingen with it. By happenstance, a scientist whom I had met in Marburg, chemist Friedrich Franz Seelig, had just received a call from Tübingen and was willing to accept me as a stipend holder to work with him at the University of Tübingen. So I found myself working at the Chemistry department there.

As a postdoc, what did you do?

The new chair was busy buying an analog computer at the time. I was sent to a course to professionally use such a machine and was then the one who worked the most with it. It was quite nice to be able to put my mind into these mutually influencing electrical variables (voltages) that one could watch performing on the screen of the oscilloscope. I liked very much this way of understanding dynamics in the same pictorial way in which also spatial intuition works. There was a kind of synchrony between this machine and my mind.

At that time, was there only one such computer?

There was another analog computer from a different firm in the department of biology, but I never saw it. These machines were the state of the art at the time.

You had this computer at the chemistry department?

Yes, and I was virtually free to use it all the time.

So you started simulating dynamical systems on an analog computer?

Yes, and instead of working with the almost infinite number of variables that one has to deal with in the origin-of-life theory, I kind of reduced the number of variables more and more, and soon discovered a 3-variable "chemical multivibrator." I constructed analogies between electronics on the one hand and chemical reactions on the other. "Liquid automata" was a name I coined and proved to be correct. It is quite counterintuitive that you can have an automaton, a discrete-state machine, which is a well stirred fluid. It indeed is a simple digital computer because the differential equations underlying both the soup and a chip are the same. The computers we are using are, in reality continuous-time differentiable dynamical systems. Automata theory is but a simplified description of these more complicated continuous systems. It was nice to get persuaded by a machine, to see the interaction of smooth dynamical variables everywhere, in chemistry and evolution, and also in electronics. I even already saw some chaotic oscillations on the oscilloscope screen in 1970, but I misinterpreted them as being noise-induced.

You were 30 years old at that time. Did you know about chaotic systems by then?

Not at all. It was only a few years later, during this on-going work with electronic systems and their analogs in chemistry, in continuation of my previous evolutionary networks, that I started to put my mind into the mutual interplay of four, and then three, variables.

When you were working on the analog computer, did you know Edward Lorenz? When you encountered the concept of chaos, did you meet him personally or did you only read his paper?

The paper was a present given to me by my friend Art Winfree in 1975. Years before, my friend Wolfgang Engelmann had introduced me to Art. In 1972, Art invited me after my first paper on the parallelism between electronics and liquid chemical systems to give a talk at Purdue University in his department. Then I saw him again 3 years later in Vienna at a conference on biochronological systems. He had attended my talk and told me afterwards that this was a little bit boring, my deductive theory of chronobiological systems, did I not do something more interesting in the wake of my liquid automata? I told him that I was thinking about a three-variable limit cycle that looks like a knot and cannot be flattened into a more or less circle-like thing but is irreducibly more complicated because it is genuinely three-dimensional. He replied that this sounded to him like chaos. I asked: What do you mean by chaos, I know the term only from traffic jams. He told me there had been a conference in Aspen, Colorado, as usual without proceedings, which he had attended, "on chaos" and that he had collected all the papers written on the subject up to that time and that he would send me a folder with them all. I thought he is a very good friend but people in America are always very positive and kind but possibly nothing will come out of this. Lo and behold, 6 weeks later I received a big folder with all the papers that had been written on chaos up to the time, most still in preprint form.

In that folder, was there a paper that stimulated you especially?

Yes, Ed Lorenz's paper of 1963, whom I would meet a year later, was there and more recent ones like Jim Yorke's and Bob May's and George Oster's, which I remember. Art wrote me explicitly that I should do him the favor of finding a chemical version of the Lorenz attractor.

What do you mean?

Art Winfree thereby made me his pupil by decree if you so wish—by forcing me to work in this area because he said he had no time to do it himself, so I should please do it in his place. Hence I felt I had the duty to try and find a method to implement the Lorenz equation in chemical rate equations, which are unlike Ed Lorenz's non-negative by definition.

So you met the Lorenz equation in these proceedings actually?

As a reprint in the folder Art had sent me.

Chapter 3
Dynamical Function Change

We enter the third session. You are now 35 years old and you just obtained a special folder from the United States that contained several papers on chaos, one of which was the Ed Lorenz paper.

Before that, I had obtained what is called a "habilitation" which is like a more advanced second thesis that has to be defended before the faculty and serves as a lifelong license to teach. It was about synthesizing arbitrarily large reaction networks with a prescribed—automata-like—behavior (well-stirred automata do exist, a somewhat paradoxical notion). Thereafter I was still without a university position, obtaining (after a year) a lucky stipend from the VW-Foundation on the topic of coupled optimizers in the abstract context of brain stimulation.

VW refers to the Volkswagen firm?

Yes, they are an important scientific sponsor. I had a joint project with Israel Lieblich. We had independently thought of coupling the brains of two animals in a mutually rewarding fashion. I had had the idea theoretically, he had planned to do it experimentally with mice, which fact greatly surprised me. Unfortunately, he fell victim to a car accident. I might otherwise have continued working theoretically in Jim Olds' footsteps on "brain stimulation reward" and might never have entered chaos theory. I was deeply interested in understanding the functional differences between human beings and other animals. I forgot to mention my previous "brain equation," which I found when I was 33 years old.

This was before you encountered chaos—so let's talk about the brain equation first.

When setting up the brain equation, I postulated that you need to have something which I called "will-o-the wisp" potentials present in the latter equation, meaning attracting or repulsive potentials that arise spontaneously in unpredictable positions in space, as an essential part of the function performed by the brain. I did not realize at the time that my potential-type brain equation is intrinsically chaotic itself, so that adding these will-o'-the-wisp potentials was absolutely redundant. So my later chaos work showed me that the brain equation was even more powerful than I had

A. Sanayei and O. E. Rössler, *Chaotic Harmony*,
Emergence, Complexity and Computation 11, DOI: 10.1007/978-3-319-06781-0_3,

thought when setting it up. The brain equation, which got derived in a step-by-step manner in a long paper presented at Trieste at a conference on the mathematics of the nervous system, thereafter did not catch-on in the biological community for a long time. Presumably because it is too mathematical, although the equation is very useful to be applied in principle in artificial intelligence, in artificial motivation and in the therapy of autism. It contains arbitrarily strong desires and is mandatory, I believe, for the realization of artificial brains.

And it is an ordinary differential equation?

Implicitly, yes. It more specifically is a potential equation much like Newton's equation but with orthogonal trajectories, and it is time dependent. It depends on your own momentary position in space while you are moving around. It also depends on the time spans elapsed since the last filling-up events for the different sub-potentials, but also on the momentary distances to the closest corresponding filling stations. It thereby yields an approximate solution to a variant of a famous mathematical problem, the "travelling-salesman problem." The same mathematical variant of the travelling salesman problem would 6 years later be described independently in a book by Gary and Johnson. While I had called it the "travelling-salesman-with-alarm-clocks problem," they called it the "decision-type travelling salesman problem." This problem possesses a lot of applications and is "NP-complete," as the technical term goes. The brain equation marks the essential difference between "behavioral biology" and "biochemical biology" in the sense of Darwin. There exists a second survival-determining environmental reality at work in biological evolution besides Darwin's slow temporal change. Namely, rapidly recurrent changes in space and time. These changes explain why we need a brain. The so derived brain equation is decoupled from the biochemical level of life and remains fully functional—indeed alive—when implemented in a robot.

Did you get any comments or feedback on this equation from other people?

Michael Conrad called the problem solved by the brain equation the "Rossler task." Some of my students got very interested. A student of my friend Kunihiko Kaneko was assigned the brain equation for a doctoral thesis after I had given a talk on it in Tokyo. But his thesis did not get finished and I have a very bad conscience for not having been helpful enough. The student was too modest to contact me directly. It can happen that one has a pupil very far away with whom one needs to have an intense interaction without knowing so. But to return, I should at this point perhaps also mention the recent book "Neosentience" written by Bill Seaman on which I was allowed to collaborate. A central motivation of the collaboration was the brain equation.

This equation has the potential to show chaotic behavior. But you did not mention this equation in your habilitation thesis?

The latter had been written a year before.

Do you want to explain the structure of the brain equation or would you rather go on with chaos?

It is a temptation to stick a bit longer with the brain equation. If it is true that the brain can be described by an equation—if you can know what a brain is for—, then you can build artificial brains which ultimately could be fitted into the head of a needle pin, as Michael Conrad predicted. He had been a student of Dirac's in Florida and happened to work in Tübingen at the time. He was a co-editor with Werner Güttinger and Mario DalCin at the Trieste conference and volume on "The Physics and Mathematics of the Nervous System" which contains the brain equation, published in the Springer Lecture Notes in Biomathematics (volume 4). The brain equation was a belated outgrowth of my discussions held with Konrad Lorenz 6 years before. The basic difference between chemical life and behavioral life—brain life—lies in a distinction which Darwin could have made and possibly did make already himself—I never checked. I always shied away from reading his books, especially the one with the intimidating title "On the Expression of Emotions in Animals and Man." One can predict chemical evolution—as Darwin already did with his theory of the "warm little pond." But while one can predict many things in chemical and biochemical evolution, the latter also depends critically on history—on many historical accidents. For example, the vertebrate eye—our own eye—is built in a way which is not quite optimal since the optical nerve, rather than coming from behind, runs right over the layer of light-sensitive cells with all its many necessary branchings—a fact which is a functional drawback. The eye of mollusks avoids this historical shortcoming and hence is less "stupid" in its otherwise convergent design. There are many accidents of history built into chemical and biochemical evolution. The brain equation, by contrast, is predictable on functional grounds alone, being independent of the biochemical history. So the theory of how the brain works is much simpler than the theory of how the biochemical machinery of life works. It was quite a surprise to me that mathematics alone suffices to predict the functional structure of brains. As a consequence, one can also understand the interaction between several functionally coupled brains, which by the way is chaotic, of course. It also allows you to understand and predict a very important point in human life—the notion of benevolence—which is an unfamiliar topic in scientific research. Benevolence means that one makes the goals of the other person one's own goals. This is a very complicated situation, mathematically speaking. You have two coupled optimizers, the one optimizer tries to optimize the optimality functional of the second, and vice versa, which gives rise to a positive feedback.

A cycle?

Yes, and this feedback then generates what I call with Detlef Linke the "person attractor." This is a favorite theory of mine which I dare mention here. I love it but it is not easy to describe to the world—it immediately invokes the question of whether the whole dream of life that we are experiencing is a benevolent gift, or not.

Is it like game theory where each wants to optimize the other's outcome?

Precisely, it is strange that this is possible, I feel. It is like heaven.

People do act like this.

Yes, someone who can act like this is called a "person." So the brain equation enables a theory of "personogenesis." It then turns out that a commonly held belief is flawed: Human beings are not persons (people) from birth—they become people by "waking up" as a person while still a toddler. This delay will make no difference for the law-makers, by the way. The toddler suddenly invents the idea that the other—usually the caretaking mother—is benevolent in the sense of wanting to make him happy. This budding suspicion of a benevolent intention being present over there towards the goals that make themselves felt in an inescapable fashion over here, leads to a cascade of events. It first gives rise to a direct optimizing response manifest in displaying tentative exaggeration of one's own pleasure—like opening one's mouth unnaturally wide up when being fed. If this improves the success as it predictably does, the concept of a cooperative intention is eventually projected into the position of the other (mother) as a simulation. This is the start of a game. The budding hunch of a benevolent intention being present over there arises in an internally simulated scenario in which the anticipated reward becomes an intended effect in a bi-directionally simulated, and in reality performed game. Niels Birbaumer is a specialist for such games in real life. Two coupled brain equations with "cognition" (an added universal simulator as it is present in higher animals) do implement game theory in a new sense. This is my favorite topic, if I am allowed to continue...

Please, do.

First, an attempt to reciprocate is started by tentatively beginning to give something to Mom despite the fact that it remains genuinely valuable—like a red ball. This giving is tried out only very hesitantly at first—with taking the valuable object back at the last moment. Tears flow when Mom overestimates what is only budding here, by her seriously accepting the playfully offered gift. Soon after this stage, a genuine "giving" act—the first deliberate present—is delivered. Sigmund Freud who first saw this budding interaction still claimed that an anal product was the first example of letting go of something good. This was a misunderstanding on his part. His description was the only rationalization by which a prude society could be offered an inkling of the deep truth of what becoming a person means. The mystery of an autonomous optimizer becoming a person in interaction is tremendous—even though this is "just" a math problem. The understanding of what happens here—an act of creation-out-of-nothing by inventing the suspicion of benevolence—is probably the most important implication of the brain equation. It presupposes that the latter is endowed with a universal simulator (cognitive map machine) which is unavoidable in evolution. Then one can look at two coupled such autonomous optimizers. One of the possible implications under exactly specifiable coupling conditions is the "personogenetic function change." Benevolence—deliberate cross optimization—is not something that can be taught. Attributing a person property to an outside place in the world, and responding reciprocally in a positive feedback, transforms the toddler into a person. This interactional implication of the brain equation entails a myriad of consequences. In particular, it allows for a causal functional treatment of autism—but this is another story. A psychiatrist from Mars would call the new functional state "the human schizophrenia." For to understand benevolence means that you split your mind up into two equally strong parts—giving and taking—when suddenly appreciating an act of giving. The Martian would note that the disease is contagious and can be exported to other species and planets and stellar systems. "Galactic export" is a neutral term.

Is there not a specific initial condition needed here? Suddenly, a new behavior appears without any special precondition?

Thank you. There indeed exist special initial conditions. You can describe the whole epigenetic occurrence in a deterministic fashion since the brain equation is deterministic. The interaction between two deterministic systems—mother and child seen as two autonomous optimizers with cognition—creates something out of nothing. If you so wish, you can call this extraordinary event a "misunderstanding." The child is possibly wrong in his simulation, picturing that the mother wants happiness to arise over here. I realize I switched from the brain equation via the add-on of a universal simulator and long-term memory directly over to the first scientific theory of benevolence...

You went ahead very fast—too fast.

One can functionally explain the whole thing in the cold language of science and mathematics. But I should have said at the outset that it is all based on a simple evolutionary fact. The human smile differs functionally from the displays of most other animals. Realizing this makes the jump ahead offered above intuitive—perhaps. So we suddenly talk about the face—the "antlitz" as Emmanuel Levinas calls it with a German word. On the human face, the smile and the laughter look virtually identical. This is the biological "trap" that underlies the "personogenetic" epigenetic function change. Before explaining further, I should perhaps say that human beings do share this coupling property with the wolf. The wolf when happy—corresponding to a high positive sum potential in the brain equation—is wagging his tail. But he also does so—a second reason for the same display—when he is bonding. That is, when the sum potential of the brain equation isn't high (happiness), but only a specific sub-potential (bonding) is activated to a high level. So, when a dog meets another dog whom he loves—or else his both beloved and respected master—, he wags his tail, too. The tail-wagging is a bonding display independent from his being happy for other reasons. The wolf or dog in this way closely resembles a human being emotionally, with the smile of happiness and the smile of affection. Every dog-owner is aware of, without being able to tell why it is, that his dog is so much engaged with the heart, that having a dog is almost as rewarding as having a child. Nevertheless there is a decisive difference. The dog is not mirror-competent. Technically speaking, the universal simulator—which is necessary to improve the performance of the brain equation if the source density in the underlying mathematical travelling salesman problem lies below a certain threshold—is not powerful enough in the wolf. It did not have to be made powerful enough in the ecological niche in which the wolf lives to make the wolf's brain "mirror-competent." This means empirically speaking that dogs cannot recognize themselves in a mirror. As is well known, dogs also cannot hold in their minds the features of a leach, or the concept of a thrown stone (which fact makes it possible to train them in a non-cruel way by throwing a stone at them whenever misbehaving, according to Lorenz). Dogs cannot project their own momentary activity into an opposing position and realize when doing so that the mirror image is bound to do the same things they do. This "mirror competence" is by no means specific to human beings, however—chimpanzees, for instance, love to look at parts of their body that they cannot see without a mirror, for example, as Wolfgang Köhler first saw in 1917.

Do most of them have a good relationship with humans?

Mirror-competent animals are always fascinating to watch, but except for elephants there rarely develops a deep friendship to a caretaker as it occurs between master and dog. I do know of a magpie—a mirror-competent social bird—and his human parent. So elephants, magpies, and presumably dolphins fulfill the precondition for becoming people if mirror competence and bonding are all that is needed. Nevertheless they usually do not fall into this "trap" as we all know. The reason is that they are not like wolves and humans with a functional overlap of the two expressions of joy in general and bonding in particular. The analogs of the smile and the laughter do not overlap as they do in the dog. Elephants are mirror-competent, as Frans de Waal could prove, so are dolphins and whales, as my friend Gottfried Mayer found out. Magpies are, as Helmut Prior found out, and so possibly are gibbons, keas, giant octopuses and perhaps even some species of mantis shrimp. Of those species of higher animals that are *both* mirror-competent *and* bonding, you can deliberately treat one young animal in such a way that, whenever he is happy and you as the bonding partner, are happy too, because you deeply like him, deliberately make the type of signaling motion that to this animal is a bonding reward (like a mother's laughter is to her toddler). It can be a bonding sound. In this way, it is predictably possible to lure a chimpanzee into becoming a 'person', or else also a little white elephant. So this is quite a "dangerous" insight compared to previous scientific and societal prejudices. Is it ethically allowed to tell humankind that its essence has not a biological but, rather, a functional origin? Which, by mathematical fact would mean that there can indeed be a white elephant who is wiser than the humans who come to ask his advice—if you treated him with the very same benevolence a human mother raises her child on. All of these dynamical predictions come ultimately from the brain equation.

If you like, we can come back to the brain equation in a later session. The next session will be on chaos and chaotic attractors.

Chapter 4
Chaos

We start the fourth session of recording the scientific autobiography. You now have received a conference folder which contained among others the Edward Lorenz paper of 1963. Next, I would like to learn how you proceeded in the chaos field, especially: how you created the chaos attractor, and how did your ideas and insights arise?

I was quite astonished that chaos exists apart from a traffic jam, and that it was possible to use this word to describe something specific in nature. I learned this from Art Winfree when he told me that my trying to find a 3-dimensional knot-shaped limit cycle in a dynamical system was very close to doing chaos theory. He gave me the homework to find a chemical reaction system which produces similar solutions as the chaotic Lorenz equation does in meteorology. The Lorenz equation has these two butterfly-like wings where the trajectory moves from one side to the other and back—it is quite complicated. Of course I assumed that this is what I had to do: try and find such a system, with non-negative variables (since chemical concentrations cannot become negative), one chemical concentration influencing the other, and there is also a back reaction, but not with two concentrations as I was acquainted with, but three. If two give you a periodic oscillation and limit cycle, as I had found 3 years before, three might give you something more complicated, like chaos. I soon realized that this was too hard for me to do—to do him the favor to find a chemical analog to the Lorenz system. Then my mind started to wander because I did not want to disappoint Art. I started to think what is the essence of 3-dimensional motions that are consistent—everywhere locally parallel—in space?

One of the earliest ideas I had was that I thought of my nose and of a rope being wrapped around my nose thereby coming closer to the tip with diminishing loop size, and that then this rope was losing hold at the tip, falling off and coming back up again towards the nose's base in a loop to be wrapped around again. I predicted that when I would try to implement a motion like this, it may be non-periodic and chaotic. 25 years later Normann Kleiner did an experiment with fluids which turned out to be chaotic and in addition, when we investigated the attractor, had exactly this rope-around-the-nose like shape. It was a concentric motion that

A. Sanayei and O. E. Rössler, *Chaotic Harmony*,
Emergence, Complexity and Computation 11, DOI: 10.1007/978-3-319-06781-0_4,

became smaller and smaller and when coming close to the middle would shoot out and then go back to the circumference, and it indeed produced chaos. In 1975, I first used a simpler way of arriving at the desired goal. I am quite sure how I hit on the decisive idea—on a way I could be absolutely certain to produce chaos in three dimensions—by using no other elements for the proof than ordinary two-variable oscillators. I should perhaps first explain that an ordinary oscillator is doing an interesting spiralling-type motion in two dimensions, which eventually reaches a circling path. Also, that Poincaré and Bendixson had realized that when you are confined to a plane and you have smooth hairlines there to describe a moving point without ever overlapping, not very many possibilities remain for you to generate a so-called flow. So in two dimensions, the most complicated thing you can have is what Poincaré called a "limit cycle." The motion hereby approaches more and more closely a closed line as the final regime, called limit cycle because as a limit of infinitely many approaching spirals, a closed curve is formed to which all the other solutions converge from the outside as well as from the inside.

Due to overlaps being forbidden for flow lines in two dimensions, it is quite easy to classify possible flows in two dimensions, as Poincaré did in 1882. I knew the essence from Bob Rosen's 1970 book on "Dynamical System Theory in Biology" which I had watched growing and which still is one of the gentlest books to introduce people to dynamical thinking. Bob was a genius in simplifying things, making them maximally clear. Coming from the Rosen School, I knew there was a system called the van-der-Pol oscillator. The van der Pol oscillator has a mathematically equivalent system which is called the Liénard oscillator. Everybody thinks it is the same system but the Liénard oscillator is much easier to understand than the van der Pol. The Liénard oscillator has a very simple underlying principle: There is a point that is moving along a one-dimensional stable manifold, a line, and suddenly this line becomes unstable and the point switches up to another one-dimensional stable manifold (there are two in all). The point can move backwards there towards a threshold, to make a switching move down again, and so forth forever, shuttling between minus one and plus one, say. This is a neat limit cycle in the sense of Poincaré. We have a letter Z embedded into a square as it were, with a motion going around the square, in a counter-clockwise direction, with the two horizontal parts of the Z slow, the vertical transitions fast. It is a kind of prototype oscillator, a maximally simple multivibrator. The point is that you can fall off a cliff, something stable suddenly stops and you fall off, but then there is another attractor from which you can then eventually come back "falling upwards" again. It is easy to understand this switching system, this cycle, which is based on what is in electronics called a "flipflop." This type, of two stable regimes between which you can continually, switch, is also called a hysteresis loop—if you are on the one side near the threshold, you have to perturb only a little bit to be suddenly free and go to the other manifold. In other words, there are two one-dimensional stable manifolds and there are two thresholds. That is the essence. I did not know at the time that this is called "singular-perturbation theory" in mathematics and is a very general principle. I suddenly realized that if you have instead of the two one-dimensional manifolds, two 2-dimensional manifolds that each contain a simple flow like an expanding spiral; then again there can be a threshold, now along a line rather than at a point, a real cliff. Now you can have the same thing again, but the switching is not between one-dimensional manifolds but between 2-dimensional planes.

That was the breakthrough. You can now arrange for everything you want on the one plane, like a spiraling-out oscillation, for example, and then you have this switching threshold. Then, after switching up there, you can have something else, for example the same oscillatory regime as below but laterally shifted—just as we had a different motion on the other level in the previous one-dimensional case. Then again there is a threshold for falling back, only this time along a line rather than at a mere point as in the case of the Z. Now we can predict—check with the aid of two sheets of paper—how the overall motion will look with both levels combined. I guessed immediately that this would give you chaos, if you did the shifts in the right fashion. I was so completely sure that this would work that I gave a talk on it in our seminar of Theoretical Chemistry and encountered a lot of ambivalent skepticism from the audience who did not want to believe me because I had made it so simple. One attendee told me that now I had lost because with this recipe on hand, he would be faster than me finding a working equation on the computer and then the result would belong to him because mere general ideas do not make a difference. Unless you have an equation put into a computer demonstrating that it works, you have nothing to present, was the argument. It belongs to the one who is faster. This looked foreign to me but it was actually quite helpful as a constructive stimulus, much like my experience 8 years before with the helpful Nobel Prize winner. During the Christmas break…

What year was it?

1975. During the Christmas break…

You found the equation?

No, I have to correct myself. I had the equation on hand before the Christmas break, but I then could write down the chapter which already contained the simulation, for I sent off the manuscript 2 days after Christmas. During the holidays I had had discussions with my father-in-law who helped me believe that what I saw in my mind was real. It turned out that the attractor generated in this way is quite pretty. You have a throng of concentric expanding spirals that fall off along a cliff in parallel, with each point then coming back towards a different point from below, because there is a strip of lines that via the cliff, and after negotiating a second upwards cliff, come back up again, re-injected. The points of the band then find themselves in a different region than when they left. This is quite complicated because it can happen that a part of the flow makes more windings before falling down the cliff again. Because of this, the distances between two neighboring trajectories, two neighboring paths, are bound to increase. This increase would never stop! This is the essence of chaos—a pair of points that is close can, because it is diverging like two windings of an expanding spiral, continue to get separated more and more and more, forever while staying in a bounded domain. This is chaos because this ongoing, mixing separation gives you something new all the time. It is miraculous: one can have an infinitely often continuing process where things are permanently separating while remaining within a closed domain.

Without crossing each other?

Without crossing each other anywhere locally, yes. So you get an amplification of neighboring distances that continues forever and is, as it turns out, of a transfinite—trans-infinite—accuracy. So the idea of chaos has to do with throwing a glance into the infinitely fine and being revealed an infinitely well-crafted structure. The latter is purely mathematical originally, but if you have an equation that is running in a computer and is doing this, it looks very nice, especially so on an analog computer. If you simulate it—you can look at it stereoscopically by making two projections—it looks very appealing in three dimensions. You can also listen to it by feeding one of the variables into an acoustic amplifier. So it became a unique experience to hear how such a chaotic attractor sounds.

In that Youtube movie of yours, the sounds from the analog computer were also recorded?

Yes—the "sound of chaos," named after Simon and Garfinkel's "Sound of Silence." The analog computer had the big advantage that one could choose, by throwing a switch, between different speeds of calculation, and so one could listen to the chaos either when it was running very slowly or when it was ten times or a hundred or a thousand times faster. The sound produced by this chaotic reality turned out to be familiar to the ear. It can be very low, a hoarse roar, or when you crank up the speed the pitch is high. Then you realize this is a vowel, for example an "a." This vowel is not completely pure because it does something new at every moment. It turns out this is familiar, because if you have a hoarse voice—as I almost had a second ago—, it is the same sound. Hoarseness is exactly the sound of chaos! Thus everybody has a perfect ear for chaos. If we switch the analog computer to a lower speed with the same motion—the same chaotic trajectory—, the sound that it makes proves familiar, too. Our flat lies close to a yellow postal mailbox where people would stop with their motorcycles from time to time to put a letter in and drive off again, and it turned out that I had often heard this very sound before—that of an idling motorcycle. Everybody has heard the sound of an idling motorcycle, but no one had listened to this chaotic system before and recognized that this dd-ddd-ddd-d is plain chaos. It is just a low-frequency chaos. So if you have a hoarse voice and let it run more slowly on a tape recorder, you can hear the sound of the motorcycle. The hoarse voice and the irregular pulsing of a motorcycle are exactly the same phenomenon. If you are waiting in your car at a red light and you look at your gear shift, you realize that the whole car is vibrating—not completely periodically, but the motor when idling is invariably chaotic. You can watch this when something is hanging from the rear-view mirror—some people have a little toy hanging there: it will make irregular swinging motions due to the chaotic forcing present. So chaos is something that is very close to everyday life. Everyone is familiar with it. It is the same as with periodic behavior, like a soprano voice. The beautiful voice of a singer is periodic. Everybody knows periodic motions and periodic sounds, but the chaotic sounds are just as ubiquitous—as this equation shows.

There are some other noises which I dare not mention that come from another opening of the human body which also turn out to be chaotic. So chaos, like periodic motions and pure sounds, is a constitutive element of everyday life. That was an interesting discovery or realization—I cannot call it discovery, it is just a plane fact which opened itself up and was a lot of fun to watch and listen to simultaneously. It also turned out to have serious applications. So in the human heart, for example, as my wife first modeled it together with the late Herbert D. Landahl, Rashevsky's close colleague. Also in the endocrine system when you have three hormones interacting in the body: They then can do this dancing either periodically or chaotically, and mostly it is the chaotic alternative, it turns out. Peter Sadowski and my wife made simultaneous recordings of several hormones dancing in such a loop.

So also in the brain's function, as recorded by the EEG?

The EEG sums the activities of millions, even billions of neurons. It is a kind of partially synchronized activity which also appears chaotic, but is "hyper-chaotic" in type. We have periodic motions, we have chaotic motions, and then we have hyperchaos, so there is a whole hierarchy of more and more complex types of behavior depending on the number of variables involved. Plain chaos is already very, very counterintuitive because of its infinitely strong amplifying power. Almost infinitely small differences are swiftly expanded into finite differences, and then this process repeats with even smaller original differences. I mean an interval that is very small is being spread out to become bigger exponentially, and then a very small sub-interval is again being amplified, so that chaos gives you an infinite power of looking into the small, the really small. It did not take us long to see that this principle had been beautifully understood already 2½ millennia ago by Anaxagoras in ancient Greece. Anaxagoras offered an explanation of the universe which started out from chaos—from the most complex—instead from the smallest and simplest, as we usually tend to think. Chaos allows you to look at the world with an eye that is focusing on infinitely fine detail, which mathematically turns out to be even transfinitely—trans infinitely—exact. Chaos is not just infinitely exact, as we try to be in science, and as computers try to approximate with finite precision. Chaos is transfinitely exact, meaning that differences play a role that are even smaller than is maximally imaginable. Its power is so strong that if two things are even infinitely close, this magnifying glass does not stop in its exponential grip and there is no limit to the accuracy that is opening itself up in this window. This is called transfinite mathematics, and Anaxagoras invented or discovered it, the transfinite thinking of chaos theory—in the twelfth fragment that we still possess of his teachings. He starts out with what he calls the "perfect mixture." This was his model of how the world originated, how the whole cosmos originated. Amazingly, he did not start out with the small and compact, as one would naturally tend to think in our own Western paradigm—take, for example, the Big Bang in which something very small is assumed to suddenly grow. He had the opposite way of thinking. He started out with a perfect mixture of everything and then wondered how out of a perfect mixture present at the beginning, the current "unmixed" state could arise. It was an *unmixing process* that he thought was the essence in cosmogony. Georg Cantor, the 19th century mathematician who invented transfinite mathematics (along with the word "trans-infinite" which he contracted into "transfinite" as mentioned) was the second person after Anaxagoras who completely understood the power of a more than infinitely exact way of thinking. This sounds crazy, and Cantor was indeed put into a mental asylum for much of his career in mathematics. Fortunately he did not get drugged—the mind-numbing drugs had not been invented yet—so that he could do his best work in a psychiatric clinic. It is only nowadays—through Mandelbrot's fractal theory—that people have learned to appreciate a transfinite complexity. The "fractals" of Benoit Mandelbrot turn out to be implications, not of chaos but of hyperchaos. So if we have one more dimension than we need for chaos—four interacting variables rather than three—, we can produce fractals—beautiful never-stopping self-similar structures. In other words, the explanation of why nature is fractal is because chaos theory gives birth to fractals when you go from chaos to hyperchaos, which is deterministic chaos acting simultaneously and independently in two mutually slanted directions. However, this next step in the ladder of complexity generated by chaos took me two more years to discover or, rather, uncover in equations—hyperchaotic flows and their cross sections.

You created that attractor, the simplest chaotic attractor which is named after you. Did you not try to find a chemical reaction based on this equation?

Actually the first system was not built according to the recipe, just alluded to—with a folding-over obtained by employing a cliff or two for the locally parallel hairlines in three-dimensional space comprising both positive and negative values. It was actually done with chemical equations that cannot have negative concentrations. The first artificially synthetized chaotic system was a chemical reaction system. For this had been the assignment Art Winfree had given me—that I should find "chemical chaos" for him.

But this was not that named attractor, it was another thing?

No, it was exactly the same kind of attractor, only the equation was more—unnecessarily—complicated, so to speak.

So the essence of this attractor was based on a chemical reaction?

Formally, yes. I first used equations that were a bit unnecessarily complicated but employed two flat 2-D flows with two cliffs. By virtue of being chemical they could not have negative concentrations.

Thereafter, you used the analog computer to simulate their behavior?

Not yet. First it had to be a digital computer because the described recipe—of having two sheets, two variables in a plane producing a diverging spiral, an unstable focus in Poincaré's terminology. This unstable focus shifted once the flow had fallen down the cliff onto another sheet, and then the flow was allowed to "fall upwards" again along another sharp cliff. This was a trial-and-error process that could have been done with two sheets of transparent paper with lines drawn onto each, with two straight thresholds acting as cliffs as described. It was mentally trivial, this result obtained by overlaying two out-spiraling flows each on a transparent sheet, with two cliff-lines drawn to mark the transition to the other sheet. This "reinjection principle" was absolutely trivial but predictably powerful. So, I made the experience that in order to find new things, you just have to combine two trivial ideas. There is a road towards success built into nature and the discovery of chaos can serve as an example.

You mean that an idea which at first doesn't seem to be important nonetheless can have a nice effect and be successful when taken seriously?

Yes: that one should take childishly simple ideas seriously, especially if two can be combined.

Now, in your attractor that is nowadays famously known as the Rossler attractor, we encounter only three parametric constants in the equation. The whole equation is based on a,b,c as it were. If you change a, b or c numerically, you will get different kinds of behavior. How did you find the right values for these constants so that the equation produces chaos? Was it by accident?

I knew beforehand that the principle works, so it was easy to find parameters that implement it. In some math books, the method I used is called "singular perturbation," and this way was indeed also used three years later independently by Christian Mira at the University of Toulouse in France. The underlying two 2-D equations to be connected via thresholds were easy to write down. What I had added, naively, if you so wish, was that I (unlike the math books) included the third, switching, equation on the same footing by numerically smoothing-out the infinitely sharp thresholds and simulating the whole thing as a three-variable system rather than stopping with the mathematically ideal two variables, plus an infinitely fast "constraint" or switch. Computers do not like such "stiff" equations. To describe these sharp transitions at the two thresholds, where suddenly things go from one 2-D plane to the other, is a bit demanding numerically since you have stiffness-generating nonlinearities. After such an almost-singular continuous flow was found and worked nicely on the computer after smoothing the corners a little bit—or rather letting the transitions take a finite instead of zero time—I became reckless, thinking it might be interesting to simplify the equations further to get rid of any "unnecessary" parts in the equations. Even the features which made the equations non-negative so as to represent realistic chemical concentrations—concentrations cannot be negative—turned out to be unnecessary in order to retain the chaotic behavior. So I could omit quite a few nonlinearities altogether and would, after simplifying the equation as much as possible by trial and error, witness the "miracle" that the maximum conceptual beauty of the originally overlaid 2 maximally simple linear 2-D flows would give way to a new maximally simple beauty of the resulting simplified 3-D flows and equations. In other words, it was possible to keep the beauty of the flow but simultaneously make the equations maximally simple. So by varying parameters systematically and putting as many of them as possible equal to zero, it was possible to find this simpler equation. It is not actually the simplest one because there is another even simpler one that I found soon after that for some reason is less well-known.

You submitted the simplified equation to a physics journal and the title was "continuous chaos"?

Yes, "An equation for continuous chaos" was the title.

Was it accepted quickly?

Yes, because it was submitted by a member from the editorial board at whose institute I had first presented the equation.

After this, did you receive special comments or ideas from other people?

Yes, I was lucky that a colleague at Stuttgart was quite taken by these ideas. Hermann Haken had started publishing on chaos a year before me as I learned later. He is a specialist in laser theory and is actually one of its inventors and had, as I then learned, found a laser equation that implements the Lorenz attractor, a task analogous to the chemical assignment Art Winfree had given me. So he was pursuing a parallel path and I was happy that there was this resonance in Stuttgart for my work in Tübingen, and I was even offered a position at the time which was a nice experience.

How did you arrive at hyperchaos? You already told us that it took two more years. Was the word "hyperchaos" created by you?

It is courtesy of Paul Rapp. When he visited me in Tübingen in late 1976, I told him that I planned to find the next higher level of complexity in 4-dimensional rather than 3-dimensional systems and I asked him whether I should, in case I succeeded call it superchaos or hyperchaos. Paul convinced me that the latter term is much better.

You mentioned already the theory of fractals. I would like to ask: Do you believe that in a chaotic attractor, we cannot see a fractal geometry but in hyperchaos, we can? What is the relation between chaos and fractal geometry?

I must say that also in ordinary chaos, you already have a fractal cross section, a fractal layering.

Before explaining further, how do you define, not mathematically but more intuitively, what a fractal is?

The term "fractal" reflects the intuition of Benoit Mandelbrot that nature—the trees…

What do you mean here by fractal?

It is an ingenious term coined by Benoit. I forgot to ask him how he had come to this word. He was in this way able to bring together all previous mathematical work done in the area. He transformed it into a common field of study and was incredibly creative. He also was the first to measure the fractal dimensionality of the cosmos—between one and two.

But again: Please, explain to me, what is the relationship between chaos and fractals?

The chaotic motion that we looked at in our minds (above) consisted of two arbitrarily close, locally parallel lines in an expanding spiral that—by getting re-injected—continued on two expanding paths forever, despite remaining in a closed finite domain all the time. Therefore, eventually, infinitely small differences get expanded with exponential rapidity into finite differences. If this is implemented in a flow—in an invertible system since differential equations are time-invertible—, then this machine generates an infinitely fine layering. If you cut through the latter you get a product of a line and a 1-dimensional Cantor set. Fractals have this property of being transfinitely—more than just infinitely—exact. This merely "layered fractal" is not yet very impressive. The fractal coastline of England—the actual coastline is not really fractal when you zoom down to the size of molecules—is a much more impressive example because it is fractal in two independent directions. You can zoom into every little detail of a coastline—or else the surrounding curve of a snowflake—and then forget that the snowflake too is ultimately made out of molecules. These "genuine" or "beautiful" fractals unlike the layered ones are not generated by chaos but by hyperchaos. So I discovered hyperchaos the same year—1977—in which Mandelbrot brought out his monumental book "The Fractal Geometry of Nature." Whereas physical nature is not truly fractal in the last instance, as far as we know. Hyperchaotic differential equations do generate genuine fractals in their cross sections and explain the beauty Benoit discovered.

The fractal dimension is not an integer?

The fractal dimension can be calculated mathematically. It is conveniently called a (non-integer) "dimension." I forgot who first invented this notion; Benoit has the history in his book. A superhuman accomplishment. The fractal dimensionality is maximally easy to calculate.

If we go to hyperchaos, we have a fractal geometry. Is there a fractal state space? What do you mean by saying a chaotic attractor is fractal?

The state space is usually continuous. I wonder whether anyone ever looked at a fractal state space—this would be a new question that is related to work done by Mohamed ElNaschie. A self-similar or self-affine fractal arises as a basin boundary in continuous hyperchaotic flows when we puncture them in two places. You can have an ordinary chaotic flow in three dimensions as we saw. Here everything is absolutely intuitive and locally parallel, and nevertheless you get this divergence of neighboring points. When we punch two holes into a chaotic flow, as is easy to do by playing with the three (in the above case) free parameters in the differential equation, one gets a layered fractal dividing line between state points—initial conditions—that either go into the one drain in the long run or into the other. If these holes are given different colors, the striped structure consists of neighboring colored layers in all sizes and sub-sizes, a product of a Cantor set and a line—or a "1-D fractal" for short. The Gumowski School—Igor Gumoswski and Christian Mira—saw this early on. As stated, any trajectory in the chaotic cross section goes eventually either to the one hole or to the other. By systematically scanning the initial conditions in the area in question by using a computer, you can color them according to where they go. What then remains is the boundary, a dividing line between red and blue. This fractal generated by ordinary chaos consists of locally parallel lines—actually there are sometimes curves that turn around which complicate the picture, requiring a third color if one is very conscientious, but this "impure case" does not detract from the main finding. The latter is a layered Cantor set. Although it is mathematically most intricate, it is not very impressive when you look at it: stripes of all sizes. This is what you get if it is only chaos that is punctured by two holes, by two separatrices invading a chaotic attractor. But if we have hyperchaos from the beginning, then the boundary between the two colored domains is a beautiful self-similar or self-affine, "genuine" fractal of the type turned into a science of its own and made famous by Mandelbrot.

So fractals are primarily not geometric structures but basin boundaries in hyperchaotic systems?

Exactly—if you believe in nature being differentiable so that flows are the description of nature as Newton proposed. Differential equations or "dynamical systems" are synonyms. Then with "hyperhyperchaos," you automatically get hyperfractals, which I herewith propose as a new term. If one adopts this term, new branchings open up as you follow this route, so it may be worth using this terminology. The fractal structures that Benoit found in his most famous beautiful example, the Mandelbrot set, belong to a special class of flows called symplectic or Hamiltonian flows. They are "structurally unstable," that is, infinitely rare in the set of all possible flows, but nature happens to be Hamiltonian on the lowest level herself—without friction. So the symplectic—or Mandelbrotian—beauty as one could call it, are the aesthetics that nature herself prefers on the lowest level. This holds true despite the fact that quantum mechanics partially irons this picture out. It is legitimate to fall into admiration when putting your mind into one of the "dragons" that Benoit described.

The Mandelbrot set is a whole universe. Unlike (or perhaps like) the real universe, it is unbounded in circumference. But Mandelbrot's universe at any rate never—absolutely never—stops in its finest details. This is something we do not know (yet?) about the real universe. Imagine something that is much finer than infinitely fine. It is hard to picture or believe. Nevertheless it was Anaxagoras who first saw this in his mind as being the basis of everything, as mentioned. Unlike what became customary to assume in later millennia, he started out with a universe that was in a maximally "mixed" state originally, and then wondered how the no longer totally mixed (partially pure) substances and structures that we ourselves see could possibly have arisen. He thereby hit on a single substance present in the perfect mixture that was "too fine to be miscible." That was the mind or Nous; Cantor's and Mandelbrot's minds.

According to this idea, the world has been created with maximally complex behavior and now it is going towards simpler behavior, and then it will perhaps go towards complex behavior again?

The mind is capable of un-mixing the infinitely fine mixture according to Anaxagoras. The mind can produce simplicity from the complex. This was Anaxagoras' definition of the mind. So the mind has a fineness to it which is absolutely unfathomable. It is a tribute to creation if someone can think of something so intimidatingly fine. I would think this is basically a religious idea on the part of Anaxagoras. It is described in his surviving fragment number 12. He speaks there of the mind, and implicitly, chaos. Although the word chaos does not appear in this fragment, even the word Tohuwabohu of the bible, took on the new meaning of "chaos" shortly after Anaxagoras. Nous and Chaos—mind and chaos—were apparently forced together because the mind "ruah" got also connected to tohuwabohu in the bible. Tohu and Bohu ("wa" means "and") obviously were older chthonic river gods originally, but no one knows for sure, it appears. Hun-Tun in Chinese mythology—the first emperor, also carried the name "Chaos," by the way. The role of the mind is played there by Shu and Hu, his friends who bored 7 holes into his head in the good intention to give him sense organs. Shu-hu means "lightening."

If we return to hyperchaos: What difference does hyperchaos make in the fractal geometry that one can see?

In ordinary chaos, when we make a cut through the flow, we get an infinitely often folded and stretched structure which has a 1-D Cantor set as a cross section, as mentioned. This is not maximally impressive as such. Nevertheless, in the case of hyperchaos we have a stretching and folding in two directions, with a "folded-towel map" as a cross section. If we again puncture two holes into this layered set (which my friend Masaya Yamaguti had independently called a "folded handkerchief"), each point in the folded area lands eventually either in the one or the other hole again. Coloring the two holes red and blue again lets every point in the area land either in the red or the blue sink. The two colors again encroach onto every point in the map—except for the zero-measure boundary which is invariant. The set of all such "genuine fractals" in the sense of their speaking to the eye has not yet been classified. The Mandelbrot set itself thrones as the simplest analytic prototype. The rest can be considered as its entourage or precursor as it were.

Precursor?

Yes. The unfathomable beauty of the Mandelbrot set is also the consequence of the existence of a very special kind of hyperchaotic flows. The flow in question is a symplectic flow, having an analytic map for its cross section. The underlying flow has never been formulated. Nobody seems to know what classes of other Mandelbrot sets exist in analytical maps. Each of them like the original one will be full of surprises, down to the infinitely (and then transfinitely) small since the structure formation never—not even in infinite time—stops. That is, these sets are "eternally self-similar" since there is no end in the smallest. It is the living illustration of the infinitely creative mind conceived of by Anaxagoras around 500 B.C. which makes itself palpable in the Mandelbrot set.

After the 1970s and 1980s, and still around the year 2000, we find a lot of papers on chaotic behavior which taken together give us a rather precise mathematical definition of what is going on and how to find new examples. This led to very nice achievements in controlling chaos and also in anti-controlling chaos—since both methods, suppression and facilitation, are important for applications. My question now in 2013 is, when we look back some years and decades in the wake of many papers published, do you think that the field of chaos has been "saturated" to date because we do not see very big achievements anymore? If not, what is the problem here?

I would see two ways to respond to your interesting question. The one has to do with the fate of any new development in science: After a while, people get used to it. Inventing the automobile is no longer necessary, only improving it—like exploiting the intrinsic chaos of the engine—which is still a topic. A new excitement is never long-lasting. So the decline in the number of papers written on chaos is absolutely normal. On the other hand, the frustration currently felt by thousands of chaos aficionados across the world about the fact that they are no longer taken as seriously as at the beginning is justified for once. I feel they anticipate that there is much more to come—whole new fields. I would bet that the people who believe that chaos is more important than public opinion currently has it, are right. Chaos is about to experience a revival, not as a mere fashion, but for promising to reveal its most beautiful and important fruit.

Is it your prediction that chaos will find a new focus in cosmology or do you have another field in mind?

Cosmology would be a side product, I would say. Everybody is justly interested in cosmology but cosmology is not the decisive target of the new chaos theory. The essential element I foresee is a return to the mind of Henri Poincaré. As mentioned before, he is the inventor of topology. He saw that a circle and a square are basically the same thing, which seems ridiculous at first but topologically they are indeed equivalent. (A pun: The name "Poincaré" means "quadratic point".) Some people have the power to see realities that afterwards everybody regards as self-evident. Steve Jobs' touchscreen idea comes to mind. Poincaré first discovered chaos, in the context of celestial motions, in the so-called 3-body problem in which three stars—or a sun and a planet and a moon—interact. This was in the wake of a famous problem, posed by the Swedish academy, of whether the solar system is stable. Poincaré had submitted a big treatise which he suddenly retracted making sure all already printed copies of the book would be destroyed at his own expenses, for he had found an error, a mistake. He realized belatedly that the complexity of this 3-body problem—it is chaotic—has an infinitely fine (no, transfinitely fine) structure. His technical invention is today still called the "homoclinic point", there are also "heteroclinic points" in his menagerie. He was the first after Anaxagoras, *pari passu* with his mathematician colleague Georg Cantor, to see the ultimately fractal reality behind the motion of three celestial bodies. So it is astrophysics which played a pioneering role in the history of chaos.

How does cosmology enter?

There is another famous fundamental physical theory at stake here, statistical thermodynamics. In the latter we also have particles that interact, just like the three particles of the 3-body problem do, but they are small particles, this time around and the science in charge of small particles interacting is called "thermodynamics"—as everybody thinks. We already encountered thermodynamics in the context of chemical evolution where it enabled "order" (life) to arise as a side effect to an overall increase of disorder. Thermodynamics is close in its spirit to chaos theory, but for a long time it was not clear how the two are related. The simplest example is three mutually repulsive particles in interaction. Nevertheless, thermodynamics is not the right theory to apply to celestial bodies in interaction because the latter are not repulsive but attractive over a long range: a different ballpark. It turns out that thermodynamics possesses a sister discipline here when the interacting particles are mutually attractive. In celestial mechanics, Poincaré saw the first example of this new field.

So celestial mechanics is at odds with thermodynamics?

Absolutely, this is what I am driving at. People thought for a long time that you need many particles in order to get the typical macroscopic behavior we are familiar with from the theory of heat—how energy gets dissipated and equidistributed in a system. Then, 127 years after John James Waterston, Yakov Sinai made the theory of billiards the deterministic fundament of thermodynamics in 1970. He proved the presence of exponential divergence of initial conditions in billiards which is the characteristic of chaos. Exactly the same thing—many particles interacting—applies in celestial mechanics. This, just like thermodynamics is chaos theory in action. It took quite a while until this parallelism became apparent. Possibly and likely, we were not the only ones to see that you can simplify this problem in a parallel manner to Sinai's method. By coming up with a 2-particle system in a T-tube it was possible to set up a smooth attractive version of Sinai's game, in which a transition towards equipartition can be observed numerically. This deterministic analog to the so-called heat death of thermodynamics, as Rudolph Clausius had called it, was obtained in Tübingen by Klaus

Sonnleitner in a high-accuracy "symplectic" simulation. By inverting the smooth repulsive potentials into smooth-attractive Newtonian ones, the opposite tendency was found—anti-equipartition. Both phenomena (equipartition and anti-equipartition) can be reduced to the theory of deterministic chaos.

What do you call it? I would like to hear the name you coined for that.

I guess what you are driving at is the word "cryodynamics." I apologize that I am so slowly progressing in my narrative.

We shall come back to cryodynamics in the next interview since it deserves more detail. Let me draw attention to the fact that we have on the one hand the deterministic nature of chaos and on the other, have the statistical approach to nature. The statistical viewpoint then is a bit like quantum mechanics. There, too, we can have 2 particles or 3 particles which are already subject to the theory, and we have the genuine statistical many-particles point of view. The deterministic viewpoint is chaos, and there is a statistical viewpoint offered by quantum mechanics. Chaos goes back to the many-body problem in celestial mechanics, and in quantum mechanics we have a many-body problem, too. Do you think quantum mechanics and chaos can be tied together? What is the meaning of quantum chaos? Or, do you believe that it does not exist?

This is an important question. It is related to Pauli's notion of "primary chance." Pauli claimed that this type of chance is different from the one we encounter in statistical mechanics where chance is basically deterministic chaos in modern parlance. However, in quantum mechanics it is something else. Pauli's "primary chance" that underlies it is not explicable with Anaxagoras' transfinite exact mathematics and with fractal theory. It rather has an element that comes from the outside of the world. There is a dice thrown from outside of the universe into the universe. The chancefulness of quantum mechanics has nothing to do with the chancefulness of chaos theory. Chaos theory is deterministic, quantum mechanics is anti-deterministic. The latter's chancefulness is not an intrinsic part of nature but appears added-on to it from the outside. In my view, Everett's interpretation is deterministic, however, Copenhagen is not. But Everett also contains a non-deterministic—or rather inexplicable—element: why a particular branch or world is chosen. This selective picking he cannot explain, so he does bring-in a metaphysical element as well. Nonetheless he has a deterministic theory, a fact which is absent in the Copenhagen version of quantum mechanics.

So we have classical chaos which is deterministic and we have quantum mechanics which is not deterministic?

It is deterministic in the hands of Everett, as far as the evolution of the wave function as a whole is concerned.

Let me make a switch, what is your opinion about "quantum chaos"?

The notion is a little bit misleading in my eyes because, strictly speaking, quantum mechanics and chaos exclude each other. My friend Joe Ford, who called himself "the prophet of chaos" in a tongue-in-cheek fashion, encountered grave difficulties within the physics community when he tried to emphasize this difference. So I learned that it is risky to even touch on the topic. But there can be no doubt that quantum mechanics is a new mechanics which contains elements that contradict chaos theory. For example, if you have quantum billiards, you only get finitely many possible patterns on a finite table, while chaos predicts infinitely many patterns. So there is a limitation to the maximum complexity that is possible within quantum mechanics.

Take a many-body problem, for example many electrons in a box that are not free but confined there, with a Coulomb potential. Now consider the corresponding Schrödinger equation. In the Schrödinger equation, the term involving the potential V reveals that the potential energy is not zero. So we obtain a very complex equation that cannot be solved analytically, since the many-body problem based on quantum mechanics and involving attraction cannot be solved analytically so far. Specifically, to simplify a bit more, we can have three electrons that are not free but exert a Coulomb potential on each other. On the other hand, the three can be looked at as forming a classical three-body problem of which we know that it has chaotic solutions. So why do we not have chaotic behavior for the electrons here?

The Schrödinger equation is a partial differential equation. Partial differential equations by definition are infinitely more complex than ordinary differential equations. Chaos as we talked about is a type of behavior described by ordinary differential equations. So if you have finitely many particles you have chaos already. Partial differential equations mean that you have infinitely many degrees of freedom from the outset. So the problem could be even more complex.

But in partial differential equations in classical physics, we can have chaos in a PDE?

Yes, even more easily than in ordinary differential equations. Also in linear partial differential equations you can have chaos. Quantum mechanics does formally involve linear partial differential equations, so from this point of view, you would expect chaos. But then we have quantization which limits the complexity.

My question here is, in the Copenhagen case, we do not have simultaneous information about a particle's position and momentum. So if you want to model three electrons which are not free but trapped in a potential trough, you could expect chaos, classical Hamiltonian chaos, with momentum plotted versus position, for example. But according to the Copenhagen interpretation, you cannot draw such a diagram because if you could, you would specify a point's momentum and position simultaneously. So based on the Copenhagen interpretation, we cannot have Hamiltonian chaos, right?

I agree. Therefore the so-called "quantum chaos" is a misnomer because the transfinite exactness of chaos is no longer valid in quantum mechanics. Everything is a little bit "wishy-washy" here unless you believe in an underlying deterministic theory which explains all quantum phenomena. There indeed are also infinitely exact features in quantum mechanics like indistinguishable particles. However, there is a big price to pay for the introduction of determinism which makes people reluctant to do so. You have to specify from the outside which one of the many universes that exist in quantum mechanics according to Everett is the one that you choose. But this is not the only price.

Do you want to say that in Everett's view, we can have quantum-based chaos again?

I was almost ready to say so but am now hesitating. Under the following condition, the answer is yes, if you assume a deterministic—that is, chaotic—world to be underlying the quantum world, then you can, from this deterministic ground, explain all quantum phenomena. Unfortunately, this deterministic underlying hypothetical world, if it exists, is inaccessible from the inside of the quantum world. So it is a purely hypothetical entity. It also has a further disadvantage: the quantum world that you derive from it, the specific quantum world, depends on the assignment of the observer. That is, you have to specify which subpart of the total universe—the deterministic total universe assumed as giving rise to the quantum world—belongs to the observer and which subpart belongs to the rest of the world. Then the whole theory formally becomes a solipsistic theory, that is, a theory which is focused on one particular observer—not just her or his body but within this body on the consciousness-carrying subsystem of the latter which might be one particular cell in the brain or even a particular electron in one cell of the brain. This is very disturbing when taken seriously because quantum mechanics thereby becomes person-centered, personalized, even solipsistic. There are experiments by which you could distinguish between these two alternatives—of either a wishy-washy quantum mechanics being applicable or else a deterministic quantum mechanics. Everything in quantum mechanics—even the Bell inequality and the "relativistic Bell experiment"—can be explained in this deterministic fashion. But people are reluctant because this is too religious as it were—to have to consider one particular experiencing mind. So it would be a catastrophe for the physics community if suddenly it turned out that you cannot describe nature without beforehand accepting that there is an "assignment" handed-out from up high to a part of the universe, namely to one particular subsystem and the attached consciousness.

After publishing the hyperchaos results, you were around 38?

Yes, I gave a paper on hyperchaos in Utah at Frank Hoppenstaedt's Summer Seminar on Applied Mathematics in Salt Lake City in 1978. This first equation contained an awful lot of nonlinear terms. It appeared in print one year later (and shows the first hyperchaotic attractor on the book-cover). In that year, also the maximally simplified—single-nonlinear-term—hyperchaos equation appeared which even could be listened to.

After that, did you continue to work on chaos?

I did. The fractal results—with the explanation of self-similarity as caused by hyperchaos—came later. There also came the proposed explanation for the famous Kaplan-Yorke conjecture with the "fat hierarchy" including "hyperfat attractors" (done with Jack Hudson). The first "explicit Poincaré recurrence" (jointly with Georg C. Hartmann) followed, and the beautiful "flare attractor s" (Hartmann's term) which grew out from the famous "Milnor attractor" to which we had been alerted by Francisco Doria. What resulted was the explanatory model of the economy as a "society of flare attractors," also with Hartmann. Finally, the new results on Hamiltonian chaos would lead to cryodynamics. All of this is, if you wish, added footnotes to chaos theory. Hereby, the "X-attractor," first called-for in 1976, is still hiding in the fourth dimension. For the empirical research program needed to find it—filming a thousand toddlers playing with play-dough—remains an unconsummated "big science" research project up to this day.

If you agree we can finish with this session and the very rewarding topic of chaos. We will continue with a new topic in the next. Thank you.

Chapter 5
Deterministic Hamiltonian and Endophysics

Last time we talked about chaos and hyperchaos, while you were around 38, let us continue from this age.

Thank you, what am I supposed to say? I always tried to dance on as many platforms as possible, mentally. The next big thing was an attempt to better understand physics; I mean non-dissipative physics. The dissipative chaos I had been concerned with is a macroscopic phenomenon by definition because it contains friction. So the friction-less part of nature came into the focus of my interest. It was a quite new experience to start to simulate Hamiltonian systems—frictionless systems—on a computer. I hit on the idea of a T-tube at the time. It was after I had struck a friendship with Joe Ford in Atlanta, and we both saw that we could entertain new and potentially controversial ideas on how two particles can interact in a frictionless manner to generate chaos and other nontrivial types of behavior. It had to do with the notion of irreversibility and whether one could fight it, that is, do something against the second law of thermodynamics. So this type of interest began in those years.

When did this idea of a T-tube come to your mind, around 38 or 39 years of age?

No, I actually was two years older already when I wrote the first paper on the T-tube. This was a transition period in which I slowly switched in this direction of trying to understand the deeper secrets of nature. There is this fundamental contradiction between chaos on the one hand and the quantum chance—the so-called primary chance of Wolfgang Pauli —on the other. The fact that you cannot predict the future in classical systems, is because you can never observe a chaotic system sharply enough to predict its future over a longer time. This claim does not stand alone. Chance in quantum mechanics is something different. Pauli called it "primary chance." Here you also cannot predict the outcome of an experiment, but in this case the idea is that this is not because nature is a chaotic system, but because something "worse" is responsible and interferes with observation. Pauli was always trying to be very sharp and to come up with the most pointed notion possible. When you are trying to understand nature, this is a very helpful strategy, it does not mean that you are an unfriendly person, it just means that you take nature seriously as an enemy, as Francis Bacon said.

A. Sanayei and O. E. Rössler, *Chaotic Harmony*,
Emergence, Complexity and Computation 11, DOI: 10.1007/978-3-319-06781-0_5,

What is the exact definition or most important feature of a T-tube?

A T-tube yields the simplest possible Hamiltonian system with nontrivial behavior. You have two particles in a T-tube, one particle in the horizontal part and one particle in the vertical part, and you can make them interact. For example, you can arrange for collisions by letting the vertical tube have an open connection to the horizontal tube. If you give both particles—or just the vertical one—a rounded upper surface, the collisions possible in the open mouth area are either gentle (barely touching) or massive. You can plot both positions in a 2-D diagram, and you get a nice billiard-table like behavior of the single point whose motion in the plane describes both motions simultaneously. This "configuration space" exactly mimics a single frictionless billiard ball moving on a 2-D table. The rounded interaction area complies also, which is almost a miracle—you always have locally symmetric in- and out-angles. So one can geometrically deduce what will happen. This was my basic idea at the time. It gives you what is called Sinai chaos.

In your paper on the T-tube configuration, did you talk about chaos in this structure as well?

Yes, exactly. There was a related third topic, namely, irreversibility. I wanted to understand whether it would not be possible to have the irreversibility of statistical mechanics in this 2-particle system, with each particle subjected to a 1-dimensional motion. I next designed a 2-particle system in a T-tube in which the two particles did not have rounded surfaces, but in which, one particle had a quadratic shape and the other, in the vertical tube, had a triangular tip with one vertical and one slanted side. It could then collide with the horizontal "brick" in two ways. If the latter is coming from the left, say it will touch on the vertical side of the triangular tip and hence be completely blocked in its motion and be repelled. But if it is coming from the other side, it will be able to push the intruding vertical particle back down along its slanted frictionless side. So coming from the one side, it would be pushed back but from the other side it would not. I wondered whether this could help understand irreversibility. It turned out that this system was more intelligent than its designer. Against all expectations, the system was not showing irreversible behavior. So this formally 4-variable system—two positions, two velocities—with its 2-dimensional configuration space that was completely transparent (I once discussed the underlying three-manifold in its phase space with my friend Gottfried Mayer-Kress in Los Alamos) was a beautiful playground.

The existence of chaos in the T-tube configuration was a separate finding?

Chaos was one of the implications of this picture. Chaos arises only if the surface of one of the two interacting particles is smoothly rounded, rather than being piecewise-straight. The configuration space then shows chaos according to a theorem of Yakov Sinai of 1970.

You were around 40 or 41?

I gave a talk on this in Sitges, Spain, when I was 42.

What was the outcome, the feedback from the audience and later, the readers?

At the conference itself there was no particular reaction, but Joe Ford became interested. He later built a little gadget which illustrated the same thing with a pellet in a little flat box. It had the form of the previous configuration space and contained a narrowing funnel if the pellet came from the one side, and a planar obstacle with a hole in it when it came from the other side. He would shake this contraption in a conference talk to demonstrate the expected preferred occupation of one of the two compartments, one smoothly narrowing up to the gap in between them, the other rectangular except for the hole leading into the opening alley. Indeed the compartment with the hole was preferentially occupied, but this was an artifact due to the presence of friction in the mechanical implementation.

On what else did you work afterwards?

This work on phase space led to an idea about the origin of quantum mechanics. More specifically, it led me back to a discovery made in 1913 in Berlin by Otto Sackur—who died a year later from a lab accident. He re-discovered Planck's constant independently in a gas, that is to say, in a radiation-free, purely mechanical context. The pertinent classical so-called Sackur-Tetrode formula for the entropy allowed me to deterministically explain—it sounds courageous—the *h* or *h*-bar of quantum mechanics, Planck's constant. So, a first possibility to explain the mysterious minimum action of nature arose.

Not capital H, *the Hamiltonian operator?*

It was indeed the little *h*. I harked back to the classical entropy formula, found by Sackur, which was later to be re-written by Tetrode in Holland as the so-called Sackur-Tetrode entropy formula. This formula is well-known and justly accepted, but it contains little *h* as a factor in the denominator. Everybody thinks this is not surprising as it describes an action—more precisely an action raised to the $6N$-th power. However, this formula had been found in complete independence from quantum mechanics. Otto Sackur had empirically re-discovered *h* in a gas here. The numerical re-discovery made was much less accurate than the original discovery of *h* made 13 years before by Max Planck. But here it was in the context of entropy measurements done in a gas, without any radiation or subatomic processes involved.

You are referring to Planck's discovery of the Planck constant?

Yes, the little *h* occurs here in the context of a Hamiltonian gas. It turns out that the Sackur-Tetrode formula contains *h*, not because *h* is put-in as a universal constant as everybody believes, but only because it is employed as a convenient unit. To give a comparison,—if you measure the circumference of the earth in kilometers, no one would claim that because the formula contains kilometers, the circumference is "quantized" in terms of kilometers in reality. In the same vein, the Sackur-Tetrode formula for the entropy is not quantized, even though it contains *h* in the denominator. *h* is only a unit of measure, but once you have recognized this, you can start to check in what kind of a classical system this unit h—I call it h^* (h-star)—comes out in such a way that it conforms with the experimentally observed value $\hbar$.

It was at that time that you wrote the paper on calculating the value of h?

Yes, but that manuscript was only published a year later in 1985, I first reported about it in Santa Fe the year before when Jonas was 1½ (we had taken him along).

That was in 1984?

When I was 44.

I had wanted to touch on that paper of how to calculate the value of Planck's constant. What is the physical origin of Planck's constant?

The Planck constant is a strange feature of nature discovered by Max Planck in 1900. It means that nature, if she is considered to be deterministic as physicists had assumed up to that moment, is described by a Hamiltonian that is somehow perturbed in the actions that occur. In a classical Hamiltonian system, there is no finite quantum of action. Max Planck had been very much astonished to find such a unit action at work in nature which he could not explain, in his analysis of the black-body radiation. Einstein would then be the first to see the deep significance of this discovery for other phenomena as well.

Do you know how Max Planck arrived at his constant?

It is a famous story, Graham Farmelo wrote a little text about it in 2002 in his book with the nice title, "It Must Be Beautiful." Max Planck himself was never quite happy with his discovery but saw that it was very important. It was a constant that obviously had fallen from heaven—just like the other big little c of Einstein. This parallelism explains why Planck struck this unprecedented alliance with a "nobody"—a young patent office clerk without a university degree yet who claimed he had found out that c is a universal constant. Planck had found his h as a universal constant and saw that both constants were completely unexpected and counterintuitive. It is always a big challenge to explain things that are absurd and "cry" for an explanation. h and c are really worth the sweat of anyone who is able to say something about them—even a nobody. Unintentionally I found myself in the lane of people who try to better understand h (and later this re-occurred with c).

How did you understand h, and how can you explain it to the reader?

The main thing that I saw is that the Sackur-Tetrode formula was right. It is beautiful, but most of all, can be derived rigorously for deterministically interacting classical particles. It is a "phase space volume"—an action raised to the $6N$-th power. This action is a direct consequence of the deterministic interaction of the particles involved. There is nothing opaque (non-transparent) about it. For example, if you change the volume of the system, or change the mass of the particles in the system, or change the temperature, then the value of this action that you get from the phase space volume—the $6N$-th power of an action—represents a system-specific unit action. It will change under the mentioned manipulations. Thus there exists in classical mechanics—a closed system of particles—what you can call a unit action. However, this classical unit action is specific for every given system. Having seen this, I wondered whether, maybe, there exists an overriding classical system in nature of which we all are a part that directly explains why (in nature) a particular value of this Sackur-Tetrode action holds true empirically. I tried this idea out by putting all the parameters I knew about the universe into the equilibrium equation on hand—to see whether one gets out Planck's value for the $h*$ of this classical system. If the universe is taken as a classical closed system—this may be a crazy idea but one can always try it out since anything rational is allowed in science—, then a unit action close to h could be hoped to come out. However, a completely wrong value was obtained. I was quite disappointed at first. Then I realized that there exists a second physical system from which there is as little escape as from the universe itself—the human brain. In principle, you can calculate the classical Sackur action of the human brain just as well—to check for the fun of it whether or not you get a value that is closer to h. So I put in all the values pertinent to the brain if you look at it as a momentarily closed classical statistical-mechanical system. This involved some simplifications, of course, as a first stab—but since electrons have the lowest mass, they contribute most strongly to the entropy and hence to $h*$. So carrying out the estimate was perhaps not an entirely nonsensical matter. I got a value equal to, I think, 0.7 times $\hbar$. Thus after putting classical numbers into a classical model, suddenly a value came out that was very close to h. It might converge even more closely if one did the same calculation more carefully with more realistic parameters. So here was some potential evidence—not very striking evidence but very surprising evidence—that maybe, h has to do with the fact that your brain is a system using which, you observe the rest of the universe. It is a quite disturbing idea and I was shocked at first to come up with this thought in 1984. But there is a nice anecdote in this context. I wrote a manuscript on this and submitted it to the *Zeitschrift für Naturforschung a,* reigned by Alfred Klemm who died this year as a centenarian. Professor Klemm asked me to name a potential referee so he could get an opinion from a trustworthy colleague, and I gave him the name of my friend Joe Ford who had seen the paper when he had visited with me in the same year and had found good words about it. I thought this was a sure way to get it published. I then got back a letter from the editor with the report by an anonymous referee that started out with the words "Dear Otto, please do not publish this paper."

That was it?

Not quite. I could not publish the paper in 1984 but the next year I could present it at a conference organized by the Hungarian Academy of Sciences on the dynamics of the brain. So it appeared in a conference proceedings volume and hence exists. Joe Ford of course was a very benevolent person who knew how dangerous it is to put out such ideas. In retrospect, I am quite grateful to him.

What did you do after that?

The next "thing" was endophysics. Endophysics, if you so wish came out of this experience even though I did not see it at the time. I was pushed towards endophysics in a roundabout way. Leibniz helped, Joe Ford and David Finkelstein helped, Werner Kutzelnigg helped, my student Jens Meier helped, Matthias Schramm helped, Kazuhisa Tomita in Kyoto helped, Roger Boscovich from the 18th century helped, and Hans Primas and Peter Weibel would do so later. It was all quite naïve at first. It had started out from trying to understand the universe from finding something universal, demonstrable inside. Later I realized that looking at the universe logically presupposes that you take seriously the position of the observer as a part of the universe. This is trivial, is it not?

This was around 1990?

It was earlier, shortly after the T-tube, it was still connected to it. The first long paper on endophysics was written in 1985 and published two years later. There was a forerunner in honor of Tomita, who had witnessed the bomb blast as a youth and died prematurely.

The collection as a book?

Peter Weibel brought the endophysics book out in 1992, the second edition in English came in 1998.

So let us now concentrate on, what is endophysics?

Endophysics is a very intuitive word like endocrinology. Endo means inside, and endophysics is that you look at physics under the proviso that you are a part of physics yourself—physics from within. The Greek name was given to me as a present by David Finkelstein after I had spoken of "Physics from within" as a new science—which was not unfamiliar to him.

If you say endophysics, this implies that we have an exophysics, too. Is working in physics not automatically a part of endophysics?

It should be, but it primarily (if perhaps naively) is rather a part of exophysics. All physics that everyone else is doing is exophysics, so it appears to me.

You mean we observe physics from outside?

This is what we pretend. We mentally place ourselves outside of the world.

If we want to understand how nature works—what is the meaning of endophysics?

Endophysics is when you acknowledge from the beginning that you are a part of the universe and try to look at the universe from within. In this way you can make a mental model of the exo- world, as we all do in science. Then we can check what differences arise between the two views. One can do this only by checking in a model, since we cannot take an outside position relative to the real universe. On the one hand, you can have a model where you are not part of the model; on the other, you can put yourself into the model as a part of it. This is greatly facilitated today since we can realize artificial minds in a computer. The long endophysics paper that was rejected by Science magazine (it had started out with the words "a new science is described") got published a few years later by John Casti in a conference book titled "Real Worlds, Artificial Minds." We can create an artificial physics within a computer model, as Edward Fredkin and Stephen Wolfram first did in a discontinuous, automata-theoretic context. The name as I said is courtesy of David Finkelstein who had held similar ideas himself. I wrote him a letter proposing that there are two kinds of physics, two basic sciences of physics, one being physics from without and the other physics from within, and he wrote back: why do you not say exophysics and endophysics? He gave me permission to use the name without quoting him. The idea soon turned out to be not totally novel. Probably, Archimedes was the first to have harbored the idea with his "give me where I may stand and I shall move the earth" (Dós moi po sto, kai kináso tan gan [Doric Greek]), or the dark Anaximander, who said that the whole is "unempirical" (ápeiron). The first modern scientist was Boscovich—Roger Joseph Boscovich—in the 18th century. He became quite famous, Maxwell liked him very much. For example Maxwell claimed in a book from 1872, the second edition of his "Theory of Heat," that the smallest parts of a body (atoms) must be "so impalpable that we cannot in any way lay hold of them." He also first described the demon—Maxwell's demon. Maxwell was very much thinking along these lines, which I did not know at the time. Boscovich wrote 2 papers in 1755 (in Latin) with very similar titles, one called "On space and time" and the other "On space and time as they are recognized by us." This is exactly the dichotomy between physics from without and physics from within. I acquired a very strong admiration for Boscovich who is becoming more and more famous again in recent years, with R. Anderton and D. Nedelkovich's book having just come out.

Endophysics means that you are a part of the model. For example, if you have an equation describing a phenomenon as a mathematical model, endophysics says that you should be one of the variables in that model?

If you try to have a complete description of this situation, you are indeed bound to introduce the observer explicitly as a part of the system that you describe. Trying to understand the system from the point of view of an internal observer is the task.

How can we involve ourselves in a mathematical model?

Only if we make a mathematical model of the universe in which we include a mathematical model of an observer who is assumed to be inside that universe. It obviously is a part of the honorable task of science to do just that. It is unavoidable, but for some reason it got postponed for centuries and millennia.

And now we can do this?

One could always do it, but in the age of the computer it is much easier. Maxwell had a much harder life when he arrived at his demon, Maxwell's demon, to explain the irreversibility in nature.

Can we consider a very simple physical system like a spring or a pendulum?

This is exophysics.

This would be exophysics—that we can model it using a Lagrangian or Hamiltonian or whatever. Does a simple system like a pendulum, have a mathematical model in this way?

Two pendula that are jointly chaotic do form an example.

Let us take just one pendulum. In the context of endophysics—how can we picture ourselves in the form of a pendulum?

Good point. We can only put a model of ourselves into a complete model universe. So endophysics is necessarily more complicated because it not just describes what you are seeing before your eyes, but the universe must contain a subsystem called "observer," and this model observer would eventually have to become a model of oneself.

You mean the observer needs a model?

Yes, the observer has to appear explicitly in the model.

Both in the model and as a model? So the observer needs a mathematical model for himself?

This model must be made a part of the bigger model of the universe.

But then every observer has to use a different model for himself?

As you say, this is the crux, that is, the dangerous part of the endophysical endeavor. This is the reason why almost no one wants to touch it because it has this solipsistic streak to it.

You mean in endophysics we must have different results because every observer needs a different model for himself. If that model interacts, for example, with the mathematical model of a pendulum, the interaction result would be different?

Yes, but of course, observers are not that different. You just have to include the brain into the physics you model. This is not as frightening as it sounds. You have to make a model of the brain in physical terms but this is not so difficult. It does not change the formalism in any way, except that you do not just look at the formalism as it is, but also at an internal boundary that is valid within the formally described world. We can take particles, for example, to model a neuron. We are made up of neurons ourselves, so the whole thing is not hopeless, since we can model a neuron, an excitable system, on a deterministic microscopic basis.

How can I model myself for example? I see a pendulum which is moving and I want to understand my interaction with the model, so I need my own model as a mathematical model, too?

Yes, but we only need an idealized observer.

How can I model myself?

There are many ways to do this, and we do not know the eventually optimal way, as of yet. We can start out by assuming that your observer was made up of particles that are mutually interacting. Even simpler, the observer could be made up of pendula—many pendula—that are interacting. Simpler still; replace the observer by a single little pendulum. It is not hard to look at what happens if the observer is replaced by a single pendulum. It turns out you get a sharply defined action, and then you get quantum mechanics out of this assumption.

Again, my question is this; we have a pendulum, a mathematical model of it, and we have an observer, for example you. You have a mathematical model for yourself, and then you let your own model interact with the pendulum model, so you get a different result than when you have just the single pendulum. Now how can you understand that different result? Again, if you want to see the different results, you should be in the external physics?

At first sight it looks as if there was the danger of an infinite regress here.

You should be simultaneously in endophysics and in exophysics?

Yes, but it turns out there is just one added level, it is not infinitely many. You can make a model of yourself as I said. Let us assume I was a little pendulum. That is the idea. The observer is replaced by a single pendulum, or by a collection of pendula. Then you can look at what happens if you make this pendulum (or these pendula) called "observer" a part of the whole system.

Again, to understand this interaction, you need to look at it from the outside? So you are simultaneously inside and outside? Would I have to tie my model to your model to understand the different interactions? What is the important result here?

Maybe the important result is that there is no blue blood among human beings, meaning that everybody has the same rights. The results come out the same whether you have a single pendulum model of the observer, or a one-hundred pendula model, or a more complicated model. As soon as you then take a zero-temperature multi-pendula observer, the system is identical to exophysics again. So it is only the finite temperature of the observer which really matters.

Can we have such an idea also in relativity? A moving frame which we watch from the outside? Or is this exophysics again?

Quantum mechanics is not yet endophysics, and relativity theory is not yet endophysics. Each of them is a theory which introduces "ad hoc" assumptions compared to a classical theory, and those assumptions need to be explained. Endophysics can, if I am not mistaken, explain both constants that underlie quantum mechanics and relativity theory. Relativity theory as such with its own strange constant, the universal constancy of the speed of light, c, in the vacuum, can be looked at from a classical point of view and can predictably be explained by classical endophysics. The so aspired-to classical view of nature as a rational machine was first identified by Buddha, I would think, and second on the same level of philosophical sophistication, by René Descartes, the two greatest minds in science in my view. Strangely, Buddha is usually not considered a scientist, but I got lost in generalities.

If we take your pendulum model, we now need to see what the salient point is. To see this, we should go outside towards exophysics?

Yes. We need to have an exo model which contains an observer. The observer may be modeled by a quite simple subsystem in order for us to get already exo results that are of endo importance.

Do you believe, that in the double pendulum, we no longer see a chaotic motion if we look at it from within?

I do. If this is true, it would be the reason why quantum mechanics contains no chaos anymore. This fact is very hard to accept by the physics community.

We do encounter quantum chaos, do we not?

"Quantum chaos" is an approximation. It always has to be written in inverted commas. So quantum chaos is not chaos, it represents an attempt to make a bridge between chaos and quantum mechanics. Quantum chaos does not have the transfinite accuracy of chaos; it has a finite resolution eventually. So it is "ugly chaos" if you wish, forgive me for using such a word. However, it can be explained by endophysics, and then the whole thing is pure chaos theory again.

You say that this endo theory can explain quantum mechanics? Which interpretation of quantum mechanics: Quantum mechanics in one world, in many worlds, or both?

It is the many worlds theory which is an implication of endophysics. This is very disquieting because, as you know, the main objection against Everett's theory of the many worlds (or many branches, as he called it) is that it is a subjectivist theory that it almost looks like solipsism. This very same reproach can be made against endophysics. It gives you an observer-specific physics, because eventually, every observer has a different position, much as in special relativity theory, not just in space but also in terms of his or her micro motions in time. Both relativity theory and quantum mechanics suddenly give the observer an overriding role, which he had not possessed before, as is well known. Exactly this fact can be explained by endophysics.

You mentioned your efforts about $\hbar$ and now also about quantum mechanics in your general theory of endophysics. Can you give some information in this context about relativity theory? Were you interested in it?

We come into gentler waters here at first sight. I was surprised to see that it is harder to explain special relativity than to explain quantum mechanics from the point of view of endophysics. If a priori speaking the world is supposed to be rational, thoroughly rational, one has to find a way to explain both quantum mechanics and relativity (special relativity and general relativity) within a purely classical view. This rationality assumption needs to be made because only the classical view is transfinitely exact, which in fact gives it a very large weight. Quantum mechanics on the one hand has inherited many features of this transfinite exactness, but on the other hand also contains this element of unpredictability which then ceases to be mysterious if the fine motions in the observer are taken into account. Relativity also has a strange observer-centeredness to it which cannot be eliminated. As it is well known, every person lives in a different frame by definition because we are shaking a little bit each with our thermal motions, as Archimedes anticipated. Some event on a distant star might be simultaneous with me at this very moment in time but not with you. So there is an element of solipsism already built-in into relativity theory. A similar element of solipsism is built into quantum mechanics. The observer is alone, in an existentialist manner. So the scandal already exists. Endophysics is the only theory which talks about these things. It touches on a taboo, but it also explains it. It gives you the hope to eventually come back to a satisfying completely rational picture of the universe and yourself.

Now two questions, maybe stupid ones: when we talk about an observer both in relativity and in the endophysics theory, should the observer be like a human being or like an animal?

Eventually the observer should be like a human person or an ape.

Should it be alive or could it be like a piece of wood? Can you look at a table as an observer?

No, it should be a dynamical entity; it should be a dynamical system. An observer at any rate is a dynamical system—the whole universe is a dynamical system, but the observer also is a dynamical system. We are familiar with the mode of action of a neuron, for example. A neuron would be the simplest version of an observer. We can make a chemical model of a neuron in terms of molecules interacting without friction. Eventually one comes up with a Hamiltonian version of a chemical observer valid on a deterministic micro level. You then suddenly have billiard balls behind the usual quantum formalism. You can apparently indeed assume that on the lowest level, the universe is built up out of point shaped billiard ball like particles. That is only an assumption, but even the Higgs theory is compatible with it.

You can consider the observer to be a flower?

An internally dynamical flower is still very complex, one would like to make it simpler.

Don't you think that a human being is more complex than a flower? Although you just modeled a human being as a pendulum?

We should use the whole brain as a part of the universe in order to find the interface. It then turns out that you can simplify the brain by taking a subsystem.

Every person is different to model, but you need some simplification or approximation to find a mathematical model for your observer. In this vein, you use a pendulum model for yourself. You use it for some other people too, but other people may have some differences to them. How do you capture these differences in your modeling in endophysics?

It is possible to make theories of observers quite independently of their chemical or nuclear chemical implementation. For example, observers living on a neutron star, made up not of chemical substances and reactions as we ourselves are, but of nuclear chemical reactions, can be conceived of as Robert Forward saw in his book, "Dragon's Egg" which he concealed as science fiction. It is an important science book. It shows people living on a neutron star's surface that are not built-up from chemical reactions as we are, but from nuclear-chemical reactions. There might be other ways like clouds—Fred Hoyle has a science-fiction story about a "Black Cloud" that is intelligent and even benevolent. All these observers would have different interfaces to the world and would have a different type of quantum mechanics, and a different type of relativity theory.

So far, we have not succeeded in considering very precise differences between observers taken as a model. For example, you can look at a model and try to interpret it. Do you have to consider yourself as a model interacting with that model? So let us consider that I am the observer. Then my model should have some differences to your own model. We do not have a way yet to consider these differences, or do we?

We can make a model of an observer. We have to do this if we want to understand the universe. There are some quite developed models of observers—I already mentioned neurons. We know that the brain is made up of many neurons. It does not matter whether we are an octopus or a vertebrate in terms of our organization. Both systems are built out of neurons in very much the same way, and so are insects. Thus we can have a model of a brain and agree that this brain would contain something like neurons. Then we can say what is a neuron—a neuron is an excitable system. What is an excitable system? It turns out that these neurons and all excitable systems are made up of particles that interact.

How do you model neurons? You need to use some theoretical physics again. For example, you need quantum mechanics to model such particles, or do you not?

You found out correctly that there is a loop in the whole idea of endophysics.

A loop, yes. If we want to understand quantum mechanics, we need a model of ourselves, so to model yourself you again use quantum mechanics. Such a loop is logically inevitable?

It is, but there is no infinite regress here as you suspect, because we can assume that the universe were classical. This is the decisive step. I mean, maybe that is nonsense, but, if we make this assumption, then we can also assume that the observer were classical. We can look at different observers, but eventually on the micro level, we come up with essentially these moving particles in boxes, more or less. Then, an important element which I only brushed upon is classical indistinguishability. If you assume that the universe is made up from only a finite number of types of particles, then they would be transfinitely equal in each class. It is a beautiful fact of nature that something called indistinguishability exists empirically. Its importance is vastly underrated. It was found again by Wolfgang Pauli, as I learned from Hans Primas who was his pupil. Indistinguishability is not just an exact, an infinitely exact, feature. It is a transfinitely exact feature. It is something which is a scandal in the good sense, an absolute scandal, because it is incompatible with any just finite or infinite accuracy. Indistinguishability is a mathematical symmetry. A mathematical symmetry is sometimes, maybe almost always, transfinitely exact. So if you have a transfintely exact Boltzmann-type universe, which one can assume, then within this transfinitely exact Boltzmann universe, you find a number of features that we are familiar with from quantum mechanics, particle-specific cells are formed amongst interacting particles. We suddenly get an analog to quantum cells. We get Pauli cells—classical Pauli cells within a classical universe. Then we can use these classical Pauli cells to build up an observer inside this classical universe. This then turns out to converge very closely with what we know of our own universe. So the express fear which we had originally that we might not be able to model anything because there is a loop in the whole thing is, as if through a miracle, solved by the existence of tranfinitely exact features (indistinguishability) in our given universe. This feature we can then transpose onto our model universe. Thus it turns out that our hope for a complete understanding of a model universe is very much supported by the notion of—the discovery of—transfinitely exact features in our own universe, present on the lowest level of particles. Pauli's discovery of the Pauli cell would be the most fundamental discovery, not just in quantum mechanics where it is a little bit of an outsider-type result, but also in classical physics. So it is again mathematics, transfinite mathematics—chaos-theoretic transfinite mathematics—that would be the basis of everything.

You want to say that endophysics is a scientific theory, not a philosophical theory. My last question on endophysics; how is it falsifiable?

If for the first time you can explain things that no one could explain before, you have a fairly good feeling. Of course, you want to falsify it again in the good Popperian tradition, but falsification also means finding a way to do so. Yet this is a system that is fairly self-consistent (I did not mention the ethical reasons that make this rationalism almost mandatory). There is a technical term for such theoretical systems: "consistent systems of insanity"—the culmination point every psychotic person tries to reach. This is also what science tries to find. As long as you have found a system, an intrinsically consistent system, you can call it "science." If no one can contradict it, then you may have to wait for years or centuries or more until a better theory is found.

I think a scientific theory has implicit in it a falsification rule. For example; any physical theory—consider quantum mechanics and the Schrödinger equation—invokes falsification implicitly. It means that if you find an experiment that does not obey the proposed equation, the latter will be falsified. It implicitly contains its own falsification rule. How does endophysics state its own falsification rule?

I mean if endophysics can explain quantum mechanics, if endophysics can explain special and general relativity, then we have made some progress also in the direction of falsifiability. Then we will have to think about the next step how to falsify this progress in a bigger context.

Another question; if special relativity will be falsified eventually, just for the sake of the argument, would then the exophysics which explains it endophysically be falsified as well?

If you have a more fundamental theory which explains another theory, and it then turns out that what it explains is false, then you must have a look whether this falsity follows also from the fundamental theory so that the latter is falsified, too.

Do you want to use the theory of endophysics as a theory of everything?

Of course, but not quite of everything. There is at least one thing that is not covered—consciousness with the Now and its colors.

Chapter 6
Einstein–Podolsky–Rosen and Everett

In the previous session we talked about endophysics. Now I want to continue with this topic by asking the question; what is Everett's main idea, and what is the relation between your own idea of endophysics and that of Everett?

I was first made aware of Everett in 1965, when I was 25 years old, by students of Carl-Friedrich von Weizsäcker who were very much interested in Everett at the time. Then I lost sight of him for almost thirty years. When I asked John Wheeler—we were already in the 1990s—how I could contact Everett, he wrote me back that Everett had passed away in 1982 (it was actually a year earlier). It was quite a disappointment to me that I could not talk to him. Everett's theory is a typical example of endophysics. The first theory of endophysics is due to Boscovich and I see Everett in the same lineage of independent minds. Everett's famous paper of 1957 in the "Review of Modern Physics" already contains the solution to the EPR-problem on its last page. It is not very well known how far-reaching his paper actually is. I later met the editor of that issue of the Review of Modern Physics in which Everett was allowed to publish his paper, and learned from Bryce DeWitt, who was quite young at the time as a guest editor, that he had asked Everett to un-clarify a little bit his equations so that Bohr would not get annoyed too much by what he had to say. It is a pity that the changes were not recorded.

Everett once picked up a hitchhiker while driving from the East coast to the West coast, so I learned from Bryce DeWitt. The passenger was a theologian by training, and they talked shop for several days during the long voyage about fundamental questions. A few weeks later, Everett learned by happenstance that this same person had committed suicide—and he was afraid that it had to do with the conversation they had. So Everett was quite polarizing a mind as far as metaphysics goes, and his theory is very metaphysical indeed. It is in the footsteps of Boscovich—an "exo type" theory. That is, it belongs into the field of endophysics which knows about this distinction. It puts the fact that the observer forms an integral part of the universe center stage in the theory. So Everett's theory is, if you so wish, a "solipsistic" theory. A person who is part of the universe shares the same wave function with everyone else, but within that wave function the actually obtained measurement result, the physical reality which appears, is

A. Sanayei and O. E. Rössler, *Chaotic Harmony*,
Emergence, Complexity and Computation 11, DOI: 10.1007/978-3-319-06781-0_6,

person-specific (private). So Everett's theory is not a "many-worlds theory" as is often claimed. Everett never said so. It can be called a "many-cuts theory." It refers to a single universe described by a unique Psi-function (as Schrödinger's wave function is called). Therefore the world is different for every observer, but not just for every observer, also for every moment in time valid for a given observer, because the observer herself or himself is continually changing in time on a micro level (as John Bell emphasized). Therefore, the universe is changing in a way the observer does not see. The observer thinks the universe has always the same past and future, but in reality even the observer is going to be split according to Everett. So this is a very Indian-type religion as it were in which a soul lives through many lives. It is a solipsistic type of theory as such. Everett was fairly well aware of this fact, although he did not publicize it. On the other hand, his theory is falsifiable and hence scientific. Therefore one has to distinguish between the Copenhagen interpretation of quantum mechanics and Everett. Niels Bohr was a very great mind but could not bring himself to acknowledging the strength of this new theory as an alternative. Everett's theory is testable—so that it is not a mere "interpretation" as is usually believed but is a falsifiable "theory" of its own. The Copenhagen interpretation of the Bohr school is also a testable theory. Experiment can show which of the two theories survives. I would be ready to bet that this will be the Everett theory and not the "Copenhagen interpretation," as the latter will be called afterwards.

What exactly is the meaning of "many cuts" in place of "many worlds"?

It is just the difference between endo and exo. When you are a part of the world, then the world becomes a subjectivist reality. This is hard to accept unless you remember how it was when you were a young child. At a very young age all these things were known to you. Later when you grew up becoming integrated into society, you would lose this insight into the strange special role one occupies in the world as an individual. The waking-up moment in life as a distinct person took place when you or I were very young, around one and a half year old, sometimes later. At that time one is very philosophical in one's mind and knows many things that later get lost because society does not know about them, so there is no reinforcement. This is one of the reasons why religion is so important. The subjectivist nature of reality and the Now—and how much one is dependent on benevolence and how much one can also give benevolence—is something young children are familiar with (and can later re-appreciate as scientists or in old age). But in between, most people are not aware of this reality, so I am almost ready to bet.

That momentary observer exists in every slice?

Yes, the state of the observer co-determines the objective reality. This fact is known in psychology, of course, but it is not something having to do with complicated things like how the brain works on the macro level. It rather has to do with how the brain is working as a physical system built out of elementary particles on the most basic level, the micro level. The brain is not at zero temperature. Which particular cut-type "world" is selected at every moment depends on the momentarily valid microscopically defined state of the observer. It is not the general noise level as such like a non-erasable temperature-induced perturbation, it is the momentarily applicable exact motion of all the particles in the observing part of the brain which co-determines how the world appears—or rather exists objectively—to the observer. It co-determines the momentary appearance of the world, both as far as the Now is concerned and, within this given Now, what micro results—quantum results—govern the rest of the world. While it is fairly easy to understand that Everett's theory refers to a "cut," this cut represents a very particular selection. It reflects the momentarily valid difference between the state of the consciousness-bearing elements of the brain and the rest of the universe. This "subtraction" or difference causes the quantum phenomena, I would say.

You mentioned an experiment which could help to decide which theory will survive. What is that experiment?

It is the "relativistic EPR experiment." It can be found in Everett's only regularly published physics paper of 1957, on the last page, but no one realized that at the time. Later a pupil of Asher Peres' in Tel-Aviv, Susan J. Feingold , was the first to my knowledge who saw the implicitly alluded-to experiment as a way to explicitly distinguish between the Copenhagen and the Everett theory, in 1978. She apparently did not publish. Later still, other people arrived at the same idea. I found this idea also in a book by Roger Penrose titled "The Emperor's New Mind" published in 1989, a few months after my own paper on the subject had been presented at Zurek's quantum conference in Santa Fe. Several other scientists later came up with the same idea including Anton Zeilinger in Vienna. There even was a proposal, made by Zeilinger to ESA in 2001, to do the experiment with existing technology. So far, the quantum community has been uneager to find out whether Copenhagen or Everett is right. Is the world really a solipsistic world? Actually, solipsism is not the right word since it implies that there is only a single mind experiencing the world. This is the opposite of what the experiment can show. Nevertheless we have here a scientific theory which explains why the world as it exists for an individual's consciousness, has the properties which it has. So it is quite frightening. The empirically decidable question posed to the experimenters of ESA represents a worked example in endophysics.

So we do not have yet this experiment realized in practice?

Unfortunately no, we haven't. It is interesting to see how the experiment can be done. You need a satellite to do it, but it is a quite easy-to-do experiment given currently available ingredients. The elements are the same as in experiments quite often performed down here on earth. In 1967, Kocher and Commins first set up the whole experiment making sure that everyone could do it. Their famous successors, including Alain Aspect, did not quote them because the inventors had not bothered to measure through all polarization angles (as trivially feasible yet later found to be necessary in the wake of the Bell inequality). The experiment allows you to look at two photons that are "correlated" when they come out of an atomic source. You measure one property on the one photon and a non-commuting other property on the other photon. Then you have measured both momentum and position (or analogs) on two mutually symmetric particles—an idea Einstein had for the first time. It is called the EPR—Einstein-Podolsky-Rosen—experiment and was first proposed in 1935.

The novel thing that Susan Feingold brought in was to measure, not just the momentum here and the position there by using mere spatial separation between two twin particles, but to use "causal separation" instead. Causal separation means that you put each measurement into a receding frame relative to the other, so that in either frame, the measurement made there is the first ("virgin") measurement. Then you have two quantum measurements which both reduce the same virgin wave function. Both do so in their own relativistic frame. The measurement made on the other side is done later—in both cases. Therefore the first measurement has determined the position of a pair of twin particles while the other has determined the momentum later which means that the momentum has been perturbed by the first measurement and is no longer a virgin experiment. But in the other frame the momentum measurement was the first and the position measurement was the perturbed one. Thus, one has *two* quantum measurement results that are both virgin—each done before the other in its own frame. This indubitable state of affairs is incompatible with the Copenhagen interpretation while it is no problem in Everett's theory where each result is specific to a quantum world of its own. According to Copenhagen, quantum mechanics would be disproved since one could measure both the virgin momentum and the virgin position of a pair of particles in defiance of the uncertainty relation. In Everett's theory, each observer lives in a quantum world of his or her own. Hence the incompatible pair of measurements known by the other observer does not exist in the world of the first, and vice versa. No matter how hard to accept, this absurdity rescues quantum mechanics in Everett's theory.

I could add that endophysics goes one step further and replaces the capital Psi of Everett for the total wave function by the capital H of the deterministic classical Hamiltonian function of the universe. So there exists a potential further layer behind Everett's theory—Boltzmann's theory. Everett did the decisive step of the voyage. Let me summarize; if the experiment is performed and reveals what no one doubts (namely, that the Bell correlations are confirmed), then this confirmation implies that the worlds of different observers cannot all be the same. If there is only one world, both the position and the momentum (or their analogs) of a pair of correlated, that is, mutually equal, particles have been measured and quantum mechanics is violated. Einstein's plan to "complete"—his word for "to kill"—quantum mechanics as represented by the Bohr school, is then vindicated. This would be one more triumphant success of Einstein. To date, there remains the loophole found by Everett. The same experimental outcome then proves that Everett's theory of quantum mechanics survives. Here each observer lives in his or her own quantum reality. Each then learns that one of the two measurements took place first in the frame to which he or she belongs, and the other came later and hence was perturbed. This "perturbed" signal cannot be the same, however, as that registered in the other spaceship in the latter's frame. The experiment then has proved that more than one quantum world exists. This loophole was discovered by Everett, and on the last page of his 1957 paper he shows that he knew this. The impending Zeilinger experiment therefore possesses maximum importance.

The question is; was the EPR proposal created to show that we are able to measure momentum and position on two correlated particles, or one of the two? In such an experiment we use two different photons made to radiate into different frames. If we measure momentum and position (or their analogs) on both, an observer in the one frame measures position and the other momentum (or an analog)?

Let me come back to Einstein's basic idea. Einstein chose an old physics textbooks experiment. If you want to measure both the position and the momentum of a particle, then you can use an experiment I learned about in high school 20 years later when Dr. German (our physics teacher) performed it for us. He took two metal toy locomotives that were mechanical twins of each other, without motors. In front of each pair of bumpers, a compressible linear spring was mounted, with an endplate. The two locomotives were pressed together so that the springs in between them got compressed, and a woolen thread was used across both to hold them compressed. The pair stood motionless. Then Dr. German approached the connecting thread with a candle-flame to burn the thread between the two stationary locomotives on the table. Both locomotives promptly accelerated symmetrically since the springs released their energy, to move with exactly the same speed in opposite directions (to come to a standstill at corresponding positions later). This exactly was Einstein's idea: to use two particles equal in all properties moving in opposite directions. Then you can measure at leisure on the one the position and on the other the momentum. In this way you have measured both the momentum and the position of each because they have exactly the same properties.

But they are two?

They are two, yes. Since you measure one thing on the one and the other thing on the other, it appears impossible that the measurement done here could disturb the measurement done there, as quantum mechanics forces you to predict.

Bohr published a paper on the same topic?

Yes, he did this in the same year, in the same journal, with the same title, so as if to confuse the rest of the world. Bohr indeed took Einstein's paper maximally seriously, dropping all other activities for six weeks to produce what he hoped to be a rebuttal. Obviously, he saw the implied danger to Copenhagen, that is, to his own interpretation of quantum mechanics.

What did he say in this paper against Einstein?

Essentially nothing new, although some details are worth looking at again. The essence of Einstein's idea remained un-disproved. Maybe I'll first give the solution before trying to explain the confusion. The solution is that Einstein did not mention with one syllable that 30 years before, he had shown that special relativity allows one to causally decouple two systems. He clearly felt he need not mention his old theory to embellish this new experiment. He was so proud he had the idea of two symmetrically equal particles that he omitted mentioning as a footnote that you could put the two particles into different frames, so that each measurement could be made in a frame of its own moving away from the other to make sure that the measurement done here comes first and the measurement done there comes later—so that each measurement comes first in its own frame. So you have two virgin particles, each undisturbed by the other measurement because that measurement comes later. Then each side reduces the wave function first. In this way, you get "too many" results compared to what is allowed in the Copenhagen interpretation. In this way the Copenhagen interpretation of quantum mechanics could be disproved. The title of Einstein's paper is…

Is quantum mechanics complete or not?

Yes: "Can quantum mechanics be completed" (literally he said; "be considered complete"). By completing quantum mechanics, he meant that measuring both position and momentum becomes possible.

Did he clarify what he means by the word "completing" used in his paper?

He was tactful enough to use the word "complete" instead of "kill" in the title. The title really means "Can quantum mechanics be disproved by measuring both position and momentum?" He clearly thought that he had done so because the experiment could be done. What he did not realize is the possibility that the first measurement would influence with superluminal speed the second measurement. This he considered improbable. He was too good-natured to expect such a trick to be played by nature—as this later proved to be the case as we know. If he had taken this possibility more seriously, he would have added: If this improbable outcome should happen, then you can—of course—separate the two experiments much more effectively by causally insulating each from the other. Then both sides taken together can predictably make a full quantum measurement of both momentum and position of both particles, as is allegedly forbidden, for each measurement is a first measurement, and the other side's measurement comes later. The captain of each spaceship learns about the other's outcome as something achieved after he had already perturbed the wave function over there by his own earlier measurement. Yet if each of the two is a virgin measurement, then you have too much information in effect—such that you have indeed laid the Copenhagen interpretation to rest. This is because you know both the undisturbed position and the undisturbed momentum of either particle.

By adding the relativistic aspect, you have actually made two different worlds?

This is the point: in each frame, the ordinary Copenhagen experiment holds true.

You have a Copenhagen experiment in each world?

Precisely, the two of them taken together amount to one too many. But in Everett's theory, this is not so: each world is a world of its own. Therefore each observer lives in an ordinary, Copenhagen-like world, so that Copenhagen survives in each of the two worlds.

In two sub-worlds?

Yes, Everett has doubled Copenhagen in this case—which fact is incompatible with Copenhagen itself but completely acceptable in Everett's theory. Therefore, Everett's theory survives while Bohr's interpretation does not. It only seemingly survives in each of the two Everett worlds.

Suppose we accept Everett's theory in place of the Copenhagen theory: would the mathematics of quantum mechanics be changed?

This is an important point. The answer is no. As we saw, in each half- or parallel world, which is a Copenhagen world if taken alone, everything is fine since one measurement came first and the other later (with the usual Bell correlations so that it must have been influenced superluminally by the first). The formalism is absolutely observed.

For the whole world?

Yes, as far as one side is concerned with the two measurements that it contains. Only interpreting both measurements as being primary and undisturbed is not allowed. But in each world, the other measurement comes later and is therefore perfectly acceptable as being perturbed by the first, only the two pairs cannot be the same in both worlds. Therefore, each observer is happy at first sight. According to Everett's theory, they can stay so because each observer lives in a different world with a different pair of results. In Everett's theory, there is no way ever to learn about the world of the other. So in each world, the second measurement result is provably disturbed by the first owing to the Bell inequality being observed. If this is the case, the other measurement cannot be the same in the other frame. Hence there must be more than one world, more than one pair of measurements, existing. In this way only Copenhagen is disproved. The individual observer could not say that there are now two worlds. Each observer has only one world. Everything therefore is okay for him or her. Each individual other person in the universe also necessarily is either in the one frame or the other: One of the two spaceship captains is the earlier one for him. Thus everything is okay in Everett's theory. But there is a terminal problem with Copenhagen. Copenhagen is "completed"—which means killed—by this experiment. Maybe this fact is co-responsible that people did not do the relativistic Bell experiment for many years.

On June 30, 2013, I was privileged to learn from Professor Zeilinger that the necessary “quantum satellite” is almost but not quite ready.

But to remind you, our question was; when we use the Schrödinger equation, we can find an exact solution for the wave function—for example in the simple case of a hydrogen atom. In Everett’s slices or branches or worlds, we can do the same calculations. Can we not add all the branches together into a complete universe?

That is the exo physical viewpoint. Yes, we can.

I mean, however, that if that is the case, both reduced equations cannot possibly work for the whole universe?

Right, the above solutions would be valid only for under-privileged inhabitants. What you ask in effect is whether there exists a privileged outside view. The latter can, unfortunately, only be construed in the mind, since in reality we do not have the whole world available to our own eyes. If you wish, this is a very religious type of thinking. As such it indirectly explains why Everett had the mentioned collision with a theologian; it was because the theologian had realized he had met a “better theologian” as far as the power of the arguments was concerned, but the two of them with their different temperaments could not come to the wonderful resolution of the seeming conflict here suggested (exo versus endo). This rationalism is acceptable both from the point of physics and of metaphysics. From the point of view of religion, there can be no problem with that, but Everett himself unfortunately thought he was an atheist, and his discussion partner took his word for that—that he now had to accept an atheistic truth. This was a misunderstanding. So a self-misunderstanding of Everett’s was responsible for the tragic outcome, it appears. Everett later made sure he would never again talk shop with a theologian about physics because this proved such a dangerous topic. He himself could not quite accept the religious (pardon me, exo) component in his whole thinking. But your question was more specific, I am afraid. I apologize that I meandered around it.

I do not understand what exactly is changed. Everett did not want to kill the Copenhagen view—or did he? He only wanted to complete it in the sense of making it round? It seems that EPR wanted to kill it, but not Everett, right? In a book just published, David Wallace from the University of Oxford quotes on the first page a few sentences from the original thesis of Everett. He says he only wants that his own quantum theory be stronger than the accepted point of view and that it can be applied to the whole world and not just to subsystems. But does it mean that he would come from one quantum world to many?

To many quantum worlds, yes.

The book says that John Wheeler omitted these sentences when he sent the manuscript to Bohr.

This could very well be. I had a long talk with John Wheeler about his pupil. It was very nice how he introduced the topic to me when we were having a long walk together in 1983 in Como, Italy. He told me unexpectedly that he had once been asked by a stranger, a young man: "Dear Professor Wheeler are you a real professor?" He had replied, "What do you mean, I would hope so." "If you are a real professor, you would accept a finished PhD dissertation written by a student without your having provided the topic." Wheeler, impressed, had replied, "Let me have a look." He found he could comply. This is the compressed story as he put it to me. In real life, it of course took longer and he had to help Everett to re-write it, and so on. Wheeler also told me that very much later, a journalist asked him, "Please, Professor, tell me what is your latest view on Everett's theory, do you still believe in it?" He had replied: Dear Mr. X, please, do grant me the favor of retracting your question. The journalist was perplexed. He got the following answer from Wheeler; "If I tried to explain to you why I no longer believe in it, it might happen that I get a relapse." It was very hard for him to survive Everett's theory, so to speak.

From the sentences of Everett which I quoted, I infer that Everett only wanted to make quantum mechanics stronger, not change it. In interpreting EPR, you are now mixing quantum mechanics and relativity theory. In this way, you try to kill Copenhagen and in the same stroke prove that we have at least two different quantum worlds?

Exactly. The whole problem of quantum mechanics is now solved; it appears to me—due to Everett. We could call it the E^2PR problem. Everett supports this interpretation on the last page of his published paper, as mentioned. There is—if I may put it that way—another Einstein to be added to Einstein-Podolsky and Rosen. This is the young Einstein of 30 years before. The E-squared-PR problem when answered can solve everything and show that Einstein was right after all. You can indeed make sure you can measure both momentum and position in a virgin form, thereby killing the Copenhagen "interpretation" if you include special relativity, Einstein's earlier brain child. At the same time, Everett's "theory" survives. Then, after Everett's theory has been proved empirically in this way, you can go one step further and replace Everett's capital-Psi wave function for the whole universe by Boltzmann's deterministic capital-H Hamiltonian function for the whole universe—arriving at a deterministic exophysical theory. Boscovich, who had this idea first if prematurely, would be resurrected. Thus, endophysics in the Boscovichian sense would win-out over Copenhagen. At the same time, it would become plain why Einstein was always right in his insisting that an irrational theory cannot be accepted, and that Copenhagen is an irrational theory. The final solution is only a tiny bit more complicated than Einstein had proposed as a first approximation. To add special relativity would not have given him any qualms, of course. Everett would find the solution that Einstein had not seen or mentioned in its trivially reiterated form, but to achieve this, Everett had been empowered by Einstein. There is this nice surviving document—a letter written by Einstein to Everett when the latter was 12 years old. Everett was almost unfairly strongly influenced by Einstein. To me, this early encouragement by Einstein gave the later Everett the strength to solve this most difficult problem in the history of physics.

You published your endophysics book in the 1990s, after the main idea had occurred to you in the 1980s when you were about 40 years old. What happened next that you would deem important to mention?

You mean I need not mention endophysics anymore now? Upon personal reflection, I indeed see a big caesura here, almost a black hole as it were. For if you wish, the biggest question had been solved. Then there, of course, came other problems to the foreground wanting to be dealt with, lesser problems. I mean special relativity and relativity theory as a whole. What is the status of relativity once you have explained quantum mechanics within an endophysical theory? Can one combine relativity theory with Boltzmann? We saw that we can combine quantum mechanics with Boltzmann.

What do you mean when quoting Boltzmann here?

With Boltzmann I refer to a deterministic universe made up from many interacting particles with Hamiltonian properties and, in the last instance, Newtonian properties.

We can have chaotic behavior on the macroscopic level of physics as we saw. How does chaos appear on the micro-level? Do we see chaos in quantum mechanics?

There is no chaos in quantum mechanics.

Why do we have no chaos in quantum mechanics?

Chaos is a limiting problem in quantum mechanics. We only have approximate chaos in quantum mechanics because we have this quantum uncertainty that acts like a finite mask. As soon as the uncertainty principle, the non-zero value of little *h*, matters for the exactitude of an experiment, quantum mechanics takes over with its built-in probabilistic finite chance. Chaos itself, which represents a deterministic chancefulness, is destroyed. This is because quantum mechanics is stronger in physics than chaos theory is—as far as we as internal citizens of the physical world can tell.

There is, for example, a simple feature in the Hamiltonian structure of classical physics, action-angle variables. You can quantize the action-angles, for example, using an idea of Sommerfeld. I mean, there exists a transformation between the Hamiltonian of the classical point of view and the Hamiltonian in quantum mechanics. Why can we not see chaotic behavior on the microscopic level?

If you are very exact in describing, we have this famous limit to observation in physics. The Poisson brackets no longer have a zero value as they had in classical mechanics, but rather a finite value proportional to h-bar in quantum mechanics, this fact is inexplicable classically. So there is this "miracle" that something that is infinitely small in classical mechanics suddenly is "bored open" as it were into a finite uncertainty. This finite uncertainty was called "primary chance" by Wolfgang Pauli. This primary chance which enters here into the world is not part of a deterministic chaos theory, it is a new type of chance, Pauli realized. He called it "primary" because it is something completely different from the deterministic chance of chaos theory. There is a "big battle" going on between the aficionados of the two different types of chance. It is a deep achievement that Pauli coined the word primary chance in order to distinguish it sharply from the classical chance of deterministic chaos which is indeed "secondary" since it disappears when one is given or assumes a privileged outside position.

Chance appears primary in quantum mechanics because the inhabitants of a deterministic world cannot possibly predict all trajectories since they do not have the privileged position of an exo observer. If you have the same universe in a computer, on the other hand, then there is no primary chance left—this is the message of chaos theory. The same message implies also that by being a part of the universe, you predictably suffer from this—to you—"primary" chance. The word here means that there is no way to explain it deterministically for you. We are here back on the same joint stroll taken with John Wheeler in Como in the spring of 1983, trying to decipher the difference between the two phenomena.

This time, it is Bohr who had to explain to a journalist, Wheeler told me: "What is quantum mechanics?" Bohr asked back: "Do you have a matchbox in your pocket?" The journalist denied. "Here is one," said Bohr and produced an empty matchbox from his pocket. "Do you have a die with you?" Bohr would produce one. "Now I am equipped to explain to you what quantum mechanics is. Please, put the die into the match box and close it without shaking it, then give it back to me very carefully." Bohr took the match box with the die in it with great care from the hand of the journalist, turned around very slowly, and then suddenly shook the match box vehemently. Then he turned around very slowly again and asked the journalist, "please, take it very carefully and open it." The journalist did. "This is quantum mechanics." The demonstration reproduced to me by Wheeler's words and gestures mean that quantum mechanics is a chancefulness that enters into the universe behind the back of the universe. Primary chance. An external metaphysical interference into the machinery of the world is quantum mechanics according to Bohr. "Primary" is taken here as a metaphysical notion. It is the equivalent of an angel that interferes. It is something that cannot possibly be explained. We could call it witchcraft as an alternative term. I once described an experiment where you could prove that quantum mechanics has the power of witchcraft according to the usual interpretation, but this is a too long story to entertain here. Nevertheless quantum mechanics really is a "scandal" in Cartesian rationalism—unless you turn to endophysics. Pauli was very much aware of this when he coined the word primary chance. It is like an injection placed into a causal universe. It is not religion, it is witchcraft, because religion lacks the element of a lack of concern.

The famous phrase of Einstein that "God does not play dice", was that coined after this explanation arrived at by Bohr, or beforehand?

I am sure, that Einstein was first with his phrase "the Lord does not play dice."

What is the meaning of that very sentence? Does it contradict the show given by Bohr or not?

Bohr was in a sense illustrating Einstein's idea. There is this empirical fact of a dice-tossing taking place behind the back of the universe in quantum mechanics.

Who is the player giving the dice and shaking it…

… behind the back of the universe according to Bohr? Yes, there is this "external chance," the "primary chance" of Pauli's which Bohr accepted. Bohr accepted that there is a witchcraft-like thing in the universe. Einstein never accepted this. He always insisted on the full phrase "Subtle is the Lord but malicious he is not." He classified Bohr's interpretation as witchcraft while insisting on a fair metaphysics himself.

Does that player or person or God who is shaking this matchbox know the number on the die?

I would say, yes. Neither Pauli nor Bohr would give the player responsibility in detail—which fact makes the whole thing a witchcraft-type idea. As soon as the one who does the shaking is not blind, the whole thing is back within the scope of a rational science—endophysics. If you have a universe in the computer, everything is transparent to you as the operator.

I mean, can the outside player know what the result of shaking the die is, and then give it to the inhabitant to open and see it?

Yes, according to rationalism.

Did Bohr believe this and simply not talk about it?

Bohr would probably have accepted Einstein's religious idea, but I am not sure. No, let me correct myself; Einstein said "the Lord does not play dice." Bohr, by contrast, assumed a blind dice-player to be at work.

I think it is important to see the common point—that Einstein and Bohr used the same simile—dice-playing—to explain quantum mechanics. Who was earlier?

Einstein was, and Bohr was influenced by Einstein. He was fighting Einstein with his performance in front of the journalist.

Maybe Einstein wanted to say that the creator can know the result. But in this context if we take up the picture, maybe the creator cannot know what the result of the dice-throwing is?

This is exactly the difference between Einstein and Bohr.

But Bohr did not talk about this—whether the creator can know the outcome of the shaking?

I understand, from a general point of view there is no difference between Bohr and Einstein: Both use a metaphorical picture to explain the quantum fluctuations.

To observers within the universe, the outcome is not known, to the creator it is known? Let us put it this way: God plays dice but he knows the result?

This would be Einstein's view. But Einstein would add that this is not playing dice in the sense of doing something which contradicts the laws of the universe, but rather is a playing of dice within the laws of the universe. This alternative is much more elegant in the sense of rationalism—which allows for chaos to exist within the universe. It can explain almost everything, so you do not need an additional outside die—the universe is full of chaotic dice already and it is in this case a deterministic ballpark. However, Bohr could not accept this, he postulated something which destroys the internal consistency of the universe—and this is called witchcraft. This ancient debate in rationalism is resurrected by the battle-of-giants between Einstein and Bohr.

The problem is; we have chaos in the macroscopic description of the world as no one doubts. But how about the microscopic description?

According to Boltzmann, we also have it in the microscopic domain. Maxwell had this insight. In his book "Theory of Heat" he wrote that the smallest parts of a body (atoms) are "so impalpable that we cannot in any way lay hold of them." Palpating means to feel the pulse. Why did he say so? He realized already that if we are a part of the universe, then there automatically exist limits to what we can observe. It is exactly the kind of limit encountered in quantum mechanics. Thus Maxwell in a sense predicted quantum mechanics in 1872, a fact which is not generally known. Who would have thought that anyone could have predicted quantum mechanics?

Let us now come to Boltzmann. When we have a collision between two molecules, then to solve this problem Boltzmann used four axioms. The fourth axiom was an assumption which he called molecular chaos?

Yes, the "hypothesis of molecular chaos."

So he used the same word—chaos—that would later become famous to say that the velocities of colliding particles are uncorrelated and independent of position in a gas?

Yes, he assumed they are deterministically fixed but postulated them to be momentarily uncorrelated. This assumption made by him is maximally ingenious I feel. It is admissible within a deterministic universe and can indeed be verified in an artificial universe implemented on a small scale in a computer.

Which year exactly did Boltzmann come up with this?

I checked it, it was in 1895 in the first volume of his lectures on gas theory.

Humankind did not know about chaos at that time?

Not under that name. Poincaré had already discovered it, and Maxwell had written about it earlier, but so of course without using the term coined much later by Jim

Yorke. Poincaré in the 1890s already saw the transfinitely exact miraculous properties of chaos theory with the marginally erotic words "homoclinic point" and "heteroclinic point" which he invented.

Can we return from chaos to Everett and endophysics?

To "Endophysics and Everett, Incorporated." Endophysics is an implication of Boltzmann's and Poincaré's transfinitely exact view. The latter can explain everything it appears. It is comforting to find Everett to be a big ally here, in endophysics. Even though he accepted the wave function, capital Psi, as being primary rather than the capital Hamiltonian H of Boltzmann's deterministic theory. (Boltzmann pronounced the capital H as "Eta", a letter of the Greek alphabet.) This difference can in my eyes be considered to be minor. I see Everett as a strong supporter of endophysics.

Everett based his many-worlds theory on the paradigm of the dividing amoeba?

He did not talk about "many worlds" but "many cuts." It all came from the amoeba, yes. The original picture in his mind was not cutting but "branching." He vividly imagined being an amoeba and dividing. The simile of this continuously dividing amoeba got cut out from the final version of his thesis. Imagine you are an amoeba—a Kafkaesque possibility. You divide, as amoebae do, and there are suddenly two amoebas looking like you. Which of the two amoebae is the original amoeba?! If you would take a poll and they could talk, each of them would predictably say "I am the original amoeba for I have divided and now have an identical twin." This goes on with every division. The basic insight of Everett's was this amoeba theory.

Chapter 7
Infinite-Transfinite, Solenoid, Mirrors

Let us continue with the scientific biographical interview. Now you are between 45 and 50 years old. Let us start out with the meaning of infinity in mathematics and physics. What is the meaning of infinity in nature?

Thank you. I would say that we all live only in the finite. Although the infinite is much more real than the finite—as I just realize. In nature, of course, the question is whether we can find anything that is infinite, either in extension or in the small. There is this interesting discovery, made around 500 B.C. by Anaxagoras, of the transfinite. Transfinite is even bigger than infinite, as Georg Cantor successfully proved. The real question when one considers the question of infinity is, whether there is anything transfinite in nature. For example, space could be transfinite in extension in principle. We made this experience with chaos theory, that the transfinite has become real. The mathematical understanding of chaos involves that the space within which the trajectories are moving, the state space, is transfinitely exact. The state space is so fine-grained that only the concept of transfinity, i.e. trans-infinity, allows one to understand chaos.

And also self-similar fractals?

Precisely. Fractals are generated by chaotic systems when the chaotic attractor is punctured. When there is a hole in the attractor—actually two are needed—the resulting basin structure forms interdigitations of all sizes, so that a fractal is formed. This fractal remains self-similar even on infinitely small scales. It is a so-called Cantor set. When you have a hyperchaotic system, however, then this basin structure is not just segmented along one-dimensional stripes but along two directions. The result is a beautiful self-similar or self-affine structure. While in the one-dimensional case the fractality is that of a Cantor set, in the 2-dimensional case it is an infinite field of blossoming flowers of all sizes without limit in the small. A Cantor set is the simplest example of transfinite infinity, discovered by Georg Cantor in the 19th century. The Mandelbrot set is a much nicer example because here, two Cantor sets are intertwined forming an infinitely, or rather transfinitely, interlaced structure in two dimensions. For example, you can take a Mandelbrot set and zoom into it and you find among other things a new Mandelbrot set with some subtly different

A. Sanayei and O. E. Rössler, *Chaotic Harmony*,
Emergence, Complexity and Computation 11, DOI: 10.1007/978-3-319-06781-0_7,

properties, as Bob Devaney so masterfully demonstrated. At all magnifications no matter how deep you zoom in, you find the same or a very similar structure again, and the question arises; what does this mean?

Julia sets?

A Julia set is the basin boundary structure valid at one point in say, the Mandelbrot set. The scandal of Mandelbrot's pictures—"scandal" taken in the positive sense of absolute novelty—is that they form the most beautiful fractals. The tradition of this type of mathematics includes many names in the wake of Georg Cantor, and some before him, like Giordano Bruno as I learned from Elisabeth von Samsonow. Anaxagoras in antiquity still throws the largest shadow.

What is the simplest example?

Let me pick out the Smale solenoid, or Smale-Urysohn solenoid, which is formed under repetitive operations with dough—if you idealize the dough so that it never starts to become grainy in the small. A ring of red dough is elongated up to twice its former length by pulling carefully on it, then the resulting longer ring is squeezed (e.g. "fat" is made drop out) so that the volume gets smaller than it was before. Now the shrunken elongated ring is wrapped up once and put back into the position of the original ring while the missing material is filled up with new—white—dough. So the red double-loop is now embedded in white dough. If you cut through it to look at its cross section, you see two round red blobs inside the round white disk of the cut. When one then continues with the stretching and squeezing and looping, one finds inside the location of each of the two red eyes of before two smaller red eyes, and so forth. This shrinking and multiplying goes on forever. One could think of a special taffy-pulling machine that does all this automatically. What is formed in the limit is a Cantor set made up from red points lying inside a white disk. The red points have zero area in the limit when the number of wrapping-ups goes to infinity. The number of red points then becomes two-to-the-infinite, if the number of wrap-ups is allowed to go to infinity. The infinity of points which applies here is a "transfinite"—trans-infinite—infinity in the sense of Georg Cantor.

What do you mean by the word transfinite?

The word transfinite was coined by Cantor, meaning trans-infinite. People usually do not know how to pronounce it and what it means, but trans-infinite is too long a name, so he preferred transfinite because this also is no longer finite for sure, but the term "transinfinite" would have been more fitting. Transinfinite—more than infinitely refined—structures were first pictured in the mind of Anaxagoras. We do not know whether there were still earlier thinkers who had had the same idea. Anaxagoras claimed that originally, everything in the whole universe including the mind had been mixed-up in a perfect mixture. This to us is a strange idea: That the universe did not start out from something simple like a point, for example, as the modern Western tradition tries to uphold with the Big Bang. The conservative alternative is to have everything start out from a state of perfect mixture. In a perfect mixture, everything has been totally mixed so you could not mix it even more perfectly. Bones and whatever you can think of, even fingernails, included, Anaxagoras says in his surviving fragment # 12. This perfect mixture appears to be an invariant state—it could have lasted forever. However, he saw a problem here; he realized that one substance was "too fine to be miscible"—the mind (Nous). The mind could not be mixed into everything else, and because of this the mind was given control over the whole mixture which according to tradition Anaxagoras called chaos. Then, after having been given control over the mixture, what would the mind do? The answer given by Anaxagoras: the mind "unmixed" the whole perfect mixture! For this reason, all things are no longer mixed together to date—because the mind could accomplish this; a very strange idea indeed.

This transfinite ancient thinking was later taken up by a small number of people in history, starting with Gregorius of Naziance who owned the last copy of Anaxagoras' book "On Nature." He must have read it through because he also used Anaxagoras' technical term "perichoresis" for the "turning around" motion of a mixing or unmixing chaotic motion, and applied it to the mixing of water and wine in a religious ritual meant to symbolize the infinite intertwinement of two persons in love—yet both without losing their identities in the process (Alexandre Ganoczy). The ritual chalice in the Roman mass of mixing water and wine stems from Gregorius, along with the dogma of the absolute equality of two persons no matter how different in power. Mathematician Hermann Weyl discovered that the tradition of thinking about the transfinitely fine, continued next with the Mutakallimun in Islam—early Islamic philosophers who taught that the word of revelation given to the prophet (the Koran) cannot be fully identified with the creator in spite of its equally divine nature. Then, Spinoza and Leibniz (who had paid a secret visit to Spinoza) took up the notion of indistinguishability in mathematics as a transfinitely exact feature. This feature according to Weyl and Primas explains the Pauli cell in quantum mechanics even though it is a purely classical concept. Forgive me that I diverged so far, but the topic is so exciting. The lovingly done chaotic mixture is too fascinating a concept—not only in physics.

In a finite volume you can have an infinity?

Right, you have infinitely many points as Cantor saw. His "continuum hypothesis" says that on the line between zero and unity, you actually have not just infinitely many points but transfinitely many.

The strangest thing is that the trajectories don't cross each other?

In state space, in dynamical systems, especially in chaos, these hairlines of a moving state point, as it follows its trajectory in time, never cross one another, except on a so-called limit point or limit cycle to which they converge. But even the limit cycle is totally alone, since every neighboring hairline—trajectory—is only approaching it, but never reaches it completely in finite time. So, it is, how shall I say, a very lonely situation to be such an attractor because nothing can ever reach you. The attractor is either privileged or underprivileged depending on one's point of view. This untouchability holds true only for a limit cycle, that is, for an attractor that is a closed line or a point. It also pertains to a trajectory filling the surface of a torus. But it does not hold for a chaotic attractor like the solenoid, about which we were talking, this originally red ring of dough that we had squeezed and then poured white dough around so that the massive ring was now white with a doubled-up narrower red ring inside (with two eyes in a cross section, and so on, to become a Cantor set, as we saw). But what remains in the limit? There is a controversy in mathematics about this, a little-known one. It rages silently between Field-medalist Steven Smale who described the solenoid mathematically on the one hand, and people who follow Urysohn in 1920s Russia who had had the opposite idea. Smale gave a mathematical theorem which states that the limit set that is formed in the limit of the solenoid is—just like a limiting point or a limit cycle—not invertible any more in its splendid isolation, even though everything else is squeezed onto it in the limit and thereby crushed out of existence. For the whole process is non-invertible in the limit.

The other side has the opposite theorem?

Precisely, they say they do not agree that everything is going to be attracted and crushed in the limit, because nothing ever arrives at the attractor in finite time. The attractor is alone as a limit set from which there is no return. This non-enviable behavior of a point attractor or limit cycle—to be lonely like a deity that attracts everything, to then crush it out of existence like an Aztek-type god (if we do not misunderstand them)—is the opposite to Anaxagoras' way of differentiable dynamical thinking. Admittedly, Anaxagoras may not have thought of attractors—as we did above in the context of the solenoid of Smale.

Smale thought that indeed the attractor—this limiting set of transfinitely many red points in the cross section of the solenoid—was an attractor in the sense of a limit cycle, namely, that everything would be crushed out of existence on approaching it. One calls this, mathematically speaking, an "invariant set." The limit point, limit cycle and attracting torus are each an invariant set so that nothing can happen to them, and everything is going to be buried on them in time, since only this limiting point or set remains invariant. However, despite this apparently reasonable logic, the solenoid attractor does not have this property: it remains transfinitely invertible (except for a zero-measure subset). Thus, paradoxically the point set of the attractor—a Cantor set—is not an invariant set; the dough continues to behave as it did on the first step, just as a fractal Mandelbrot set does, forming a never ending abyss for the pondering mind. Abyss is not the right word here, another term could be that there is a never ending love. One could say that it is an aesthetic idea to see such accuracy in the mind of a person or in nature. Anaxagoras presumably is correct that this holds true—not just in a mathematical universe but also for the real cosmos if tried to be understood by the mind. The mind would remain and the mind would be given the power to un-mix the perfect mixture. These very unusual ways of thinking can perhaps be fully understood only by people who have played with chaotic attractors and play-dough for a long time—thereby learning to have no qualms about putting their minds into such a transfinitely exact, never stopping structure of the Mandelbrot variety. This is more beautiful, I feel, than a finite view of the universe—or mathematics—can ever be.

You just mentioned the word "love." What is the exact meaning of love in your view? Do you believe that it is possible to define love, or do you really think it is something strange like color or the Now which cannot be defined in physics?

Thank you for the question. I agree that color and the Now already fall totally out of the frame. Love is, being fascinated by something impossible, but nonetheless true. Love also has this element of...

Infinity?

Of infinite creativity and trust. Trust, being part of love, has the same structure. Even the trust in one's own mind as displayed by Anaxagoras has the same structure: not to be afraid of thinking of a beauty that never stops. I appreciate the notion of love being mentioned because this type of thinking is not just the thinking of a few people in the history of science or philosophy—Anaxagoras, Cusanus, Bruno, Cantor, Mandelbrot—but it is also something that every human being knows. For she or he invented it as a young child. I claim that this sort of pictorial mathematics has one specialist: the young child. Young children at a certain age make a sudden transition in their minds from a way of functioning which, if you so wish, is trivial—I would call it "animal-like" even though I do not think negatively in any way about animals. The young children use the powers of their animal-type minds to invent something which did not exist before or need not have existed before. It is the invention of the hypothesis of, I could say, deliberately confirmed trust. Or I could say love—it is the same thing. When you trust totally, this trust means that you accept that there is love, only love is worthy to be trusted fully, and only fully trusting is love.

Next question: If we have trust, there can be different degrees of that. Should the definition not always contain this degree?

A very important point, we know that everything can be additive and usually is additive in nature. Most everything comes in finite portions. Trust makes an exception here. How come that a young child, of but one-and-a-half years of age for example, can come up with the idea of something infinitely or transfinitely trustworthy existing—namely, the love of the mother. Then the child switches—in an un-researched dynamics which usually takes place in the playroom—into a new type of dialog with the kindergarten nurse or the mother or father of an infinite solidity. One observes an interactional switching towards this crazy type of transfinite thinking, of trust and reciprocating, that we discussed above. It has to do with a type of reflection that is like a ray passing indefinitely often between two mirrors. This awakening has to do with the smile. Why do I mention smile theory here? It is because the smile of human beings can be investigated and understood by science. In science, one looks at human beings as an animal species and tries finding out what is different with this species compared to other species. One then finds something quite intimidating: The smile is the decisive trait of a particular ape. I call him the parent feeding ape. If you are a biologist coming from Mars, you find that there exist different types of apes on our planet, and one of them has a very strange functional description. These apes are characterized by the fact that the young children love to feed their parents or to do other favors to them on a reciprocal basis. The young child wants the mother to eat this and to smell it, to taste it and to love it. Why? Why should a child feed Mom? In a biologist's eyes this is a so-called "lethal factor" because it is counter-adaptive. It is functional similar to parents eating their young—pedophagy—as certain spiders do, a so-called lethal factor. It is predictably going to be eliminated by natural selection since it does not make sense to first produce offspring and then to kill it. But we were concerned with the question of degree.

Once in the Zoo in Berlin—I shall not forget it—we were watching wolves and in front of me, there was a father with a young son also watching the wolves, and I watched the two with waxing interest. The young child had a sweet in his hand, reached up and put it into his father's mouth, asking "good?!" Apparently the father was meant to find it good, and it was meant as a gift from the child. I was intrigued enough to follow the two zoo visitors on the way out and asked the father, please let me ask how old your child is, and he said 18 months. So a child of 18 months can put a sweet into another's mouth and ask for confirmation that he enjoys it. That is benevolence. Benevolence is pathological—as we saw, parent-feeding is pathological. There are ways to explain the phylogeny of this human trait. It is not an innate trait as such, but rather a feature acquired in an interactional function change, a so-called "hard bifurcation." The epigenetic accident happens because human beings have a certain property on their faces which other animals lack. Only wolves and dogs have a similar property, of lesser degree, mediated not by the face but by the tail. Happiness and affection have the same expression. Tail-wagging in the wolf, smile in the human. It can be simulated in two coupled brain equations. Innate releasing mechanisms lead to a bonding reward. This occurs under two unrelated biological conditions here. One is a display of bonding shown by the partner, the other is a display of general happiness. Happiness is often displayed in biology. The Crislers filmed "Arctic Wild" for Walt Disney in the 1950s and showed that the joy of the young while being fed is the main motivation for a night-long frustrating hunt of a parent wolf. But why should parents have a special display for their own happiness rewarding the young? This does not make biological sense. In wolves and in humans, this is the case because happiness displayed, looks like bonding. The radiating face of the happy parent is the same as in bonding so that the young predictably starts to bring sacrifices. This can become overcritical, triggering a total change of behavior, a function change in the sense of Robert Rosen.

Do we have any model explaining this behavior mathematically?

The brain equation eventually does this. But there is a direct evolutionary explanation for the phenomenon itself, as I was told by Konrad Lorenz. Huxley, one of the famous Huxleys…

The Huxley of Hodgkin and Huxley?

Neither he nor Aldous nor Darwin's friend Thomas Henry Huxley, but the latter's grandson, Julian Huxley who was a senior friend of Lorenz. Julian Huxley studied the underlying phenomenon in different animal species under the name "evolutionary ritualization." How come that the bonding behavior that exists in social species sometimes disappears in evolution and then reappears later in a new niche? In between, the species is no longer as social as its ancestors were, then, being social is "rediscovered" in a new environment. Nature has to invent a new signal for bonding in each case. Usually, signals that already exist are taken over being given a new inherited meaning. "Ritualization" means that a signal which used to have a specific meaning is slightly altered—usually stereotyped, hence the name ritualization– to carry a new meaning. The best example, perhaps, is the African wild dog, *Lycaon pictus*. This is a type of hyenas—they are related to hyenas but form a different special species living in Southern Africa, for example. These distant relatives of the wolf are maximally social. They actively cooperate, the young are being pampered by every member of the pack, and they are very close showing bonding behavior very often. How do they show their bonding affection? They show it not by tail-wagging, they show it by mounting. Mounting has hereby been stereotyped—ritualized. They all mount, the females like the males, but it does not mean sexuality. Rather, the ritualized behavior is an expression of bonding. The selection pressure was so strong that the females evolved a special organ called a "pseudopenis" for the purpose. Even in a close relative to the human species, the bonobo, also a sexual type of contact is used for the purpose of bonding, orgasm included, which is generally misunderstood by the public and the professionals alike. They smile with the lower part of their body, so to speak. It looks as if they were always having sex along with the right expressions, and humans tend to get jealous. But this is a misinterpretation. It is an example of Huxley ritualization and signals bonding. In the same vein, one can now explain that the smile of happiness of the satiated infant got ritualized into the bonding signal for a primate species. Just by accident it was not a sexual gesture this time around but another innate gesture pre-existing in the species in question (the display of happiness shown by the satiated baby). So the cheerful smile became a bonding signal for a particular species of apes which happens to be the human one. The "same" thing happened in the evolution of the wolf, only with the signal of submission and bonding being ritualized into acquiring a second meaning, that of displayed happiness by the young. Here feeding seems to have required a stronger emphasis in a previous phase of evolution. Watching evolution in action is much like watching taffy-pulling. At any rate it somehow happened that human beings became close functional relatives to the wolf because in wolves, too, bonding and happiness displayed, have converged into one single action pattern. In either case, the manifest outcome consists in nonsensical built-in rites. In the wolf, it has no major unexpected consequences. It only explains why humans hit on their best friend—the dog. But why then is the dog not human in every essential functional respect? Humans do accept the dog as a member of the family, but dogs have some features which separate them functionally from humans. It is not that they have no fingers and smile with the tail, which would not make for a problem: They do not display this transition towards believing in infinity that we were talking about. They do not believe in infinite love, at least I dare claim this even though Lorenz wrote in a book that there never was a greater love in the universe than that shown to him by his female dog "Stasi."

I met as a child a theologian who loved his dog, which had developed cancer in old age, so much that he divorced his wife to live with his dog even though the dog would have to die soon. This theologian—his first name was also Otto—was convinced that this dog was the essence of his life. But with humans, the same *amour fou* thing not only can, but must happen owing to the positive feedback caused by happiness acquiring a radically new level. We saw that when the child is laughing out of happiness, the mother is laughing out of bonding. The good feeling of the child acts as a bonding reward, as it does in many species, so mother is also happy. She then also smiles. A positive feedback—a bonding bout—occurs between the smile of mother which is a smile of happiness and the bonding of the child, who is rewarded by the mother's smile looking like a bonding smile. So the child tries to reinforce the mother's smile of happiness because it looks like bonding. This is pathological biologically speaking, just as in the wolf where this cross feedback predictably also occurs. However, in the wolf the consequence does not make for much of a difference. This occurs only in humans. The reason is that humans have a bigger brain. In this respect, humans again are not alone; quite a few species have a similarly high simulational capability in their built-in flight simulator, as humans have. There is a name for it: mirror competence. Many apes can recognize themselves in a mirror just like human beings, but dogs and wolves cannot do that, at least there is no recorded example of a dog recognizing himself in a mirror (although a friend of mine, Boris Schapiro claims to have an observation to the contrary).

They think that they see another dog?

Exactly, and they learn to ignore that boringly synchronized apparition after a while. Nevertheless they can get used to using a mirror productively: If someone is coming up the turning staircase and there is a mirror half way up, the dog efficiently uses the mirror to anticipate the person coming up. No wonder. But to use the mirror in order to understand something about one's own position in space or one's own teeth, for example, represents a different level of using the mirror. This "mirror competence" shown by chimpanzees for example, requires more hardware in the brain. The image processing unit—the universal simulator—has to be strong enough that, in the simulation, one can put oneself into a different position and look from that position back onto one's own momentary position. Only in this way can one be mirror competent and scratch an unfeelable blemish, for example. It came as a big surprise that certain birds can make that transposition, too. Helmut Prior discovered this 12 years ago in the magpie. Now we have brought together two things to explain infinity, right? One was the smile; the other was mirror competence, two quite unexpected bedfellows. To understand Georg Cantor, however, we need both of them.

The consequence of the bisociation—the putting together of two at first unrelated topics according to Arthur Koestler—is not easy to describe. The child, seeing the mother smile out of sheer happiness, is rewarded, but this not all. At the same time, the child is being made happy by mother giving him something, a fruit to savor, for example. But then he will someday suddenly acquire the suspicion that mother over there is anticipating the joy felt on his part. The mirror competence alone is not sufficient for this transposition completely into the other. It must be combined with a positive feedback. The suspicion that what is being done over there, an act which one can understand through mirror competence, is motivated by the desire to see me be happy here. This suspicion—a creation of the simulating mind captured out of the blue, since this is what simulators are meant for—is called benevolence, or "suspicion of benevolence." The child makes the strange invention of asking himself or herself the question: does the smile perhaps mean that the happiness experienced here, is something that is desired over there? An at first sight almost erotic suspicion, modern skepticism would say, but it has only to do with the "polymorphically perverse nature of the female," as Freud lectured which was total nonsense. It is the purest form of genuine love that exists, with the wish that there should exist happiness over there, too, which the child immediately produces in response. The whole thing develops in a positive feedback. A whiff that this good feeling of a marvelously sweet apple is cherished over here stands not alone. The child in response shows more happiness—exaggerates the appreciation—by opening the mouth artificially wide, for example. I knew one little child who exaggerated how good something tasted, he was about 13 months old. This was the beginning of this nascent process. Once the child has conjured up the suspicion of benevolence, he starts checking on that. The child will give something good to the parent but then take it back at the last second, to the bewilderment of the parent. When you take it, there will be tears because his type of swimming was not yet secure. Then, there suddenly is the abrupt transition towards giving, with an almost violent not-taking-it-back. This latter behavior was recently observed by Willie Smits on an Orangutan friend in the wild, by the way; the observation makes one tremble. The next stage is that the child will ask "good?!" as we saw. This stands at the end of an unlimited positive feedback, much as in chaos with its reinjection principle. The attractor formed here was given the name "person attractor" by my late friend Detlef Linke.

Does this relationship remain many years after, or not?

It causes a function change, so this is called, an irreversible transition.

But some children at age 18–20 want to be completely isolated from their parents and they do not have such a relationship with them anymore.

They are adults then, they are mature enough to have children themselves so to speak. I remember when I was 16—I never related that story before—my grandmother came visiting us from abroad, and I liked her. But I could not explain to myself why I should love her, very strange. As a child, I loved her, of course. But in this case I realized that I could not find any reason for that anymore. It is almost a Dostojewskian story. I have the suspicion that adults are no longer full human beings. But this transfinite positive feedback that we looked at in the solenoid of Smale with more and more red dots appearing on the cross section is also…

Is it a positive feedback that makes the system unstable?

Yes, indeed. It is a positive feedback which does not stop and which eventually allows you to enter the domain of the transfinite, this Anaxagorean mathematics.

From the dynamical-systems point of view, we do not have a limit cycle because we have a positive feedback which makes the system not go to one point or a simple oscillation and stay there?

Yes, the mind is always jumping and is trying to conquer the universe, to include everything. Benevolence has inherited this intrinsic feature. The invention of benevolence suspected, and then trying to make mother happy is then, in the next stage, followed by the toddler asking "questions"—an unheard-of thing in the rest of biology: What do you want? This is absolutely pathological from the point of view of biology, when uttered by a child. The concept of benevolence is transcendental, but also transfinite. It causes a function change in the brain, in the use of the machine code employed in the brain. Theologians might be very interested in this. The brain equation—if you believe in the brain equation—remains unchanged in the bifurcation on hand, the personogenetic bifurcation. You can use the brain equation with its universal simulator in explaining both the Smale-type and the smile-type functioning and its consequences. They are very close mathematically speaking. I hope that Steve Smale will laugh about me when reading this and will forgive me that I adduced child theory to encourage him to re-ponder his claim that the solenoid has a noninvertible invariant set as its attractor. There exist sets here—limit sets—that are as open as the universe, as unlimitedly infinite as the universe. It is nice to be allowed to talk about chaos theory and its secrets and the secrets of the smile in the same mathematical context. For we never left mathematics, did we?

You have mentioned these ideas in your book titled Benevolence?

I do not have a book with this short title, but it would be a beautiful idea to write a book just on benevolence in its own right.

Did you not participate in writing a book on the art of intelligent machines, with the word "benevolence" in the title?

Yes, the book of Bill Seaman and mine is titled "Neosentience…"

…with benevolence in the subtitle…

… "the Benevolence Engine," you are absolutely right. I learned here that one can be much more outspoken or dangerous when telling the truth in front of a camera than when writing a book. It is not so easy to make yourself understood in writing because that is a more indirect medium.

Next time I plan to take you on a ride through cosmology.

Great!

Chapter 8
Cosmos, T-Tube System, Poincaré and Boltzmann

You are around 50 years old in our story. Maybe it is a good idea to talk in this session and perhaps in two sessions about the connection between cosmology and your ideas?

Cosmology is a much simpler topic than is endophysics because one can treat the cosmos from a detached outside perspective. This is what one would like to do because all directly visible objects in the cosmos are very macroscopic. On the other hand, there are these quantum phenomena of photons traversing the cosmos. In particular we have this beautiful finding of 1929 by Hubble—the Hubble law—which Father Lemaitre had qualitatively anticipated 2 years before and hypothetically explained with a Doppler effect caused by expansion in the sense of the Friedmann solution to the Einstein equation. This proposed explanation dominates the public opinion on the planet since 1929—the "Big Bang" as it would be called later in 1949 by Fred Hoyle. The Big Bang is a very beautiful concept—very deep also because of its philosophical and theological connotations. On the other hand, this concept also represents a big challenge.

When I was still a medical student, I already developed the idea that there was bound to exist another explanation for the cosmological redshift, the change in color and energy of the light emitted by stars as a function of distance. The stars appear to be receding since their light is reduced in its frequency in a distance-dependent fashion—in a linear fashion proportional to distance. While this is a wonderful finding, its proposed explanation automatically meets with skepticism from the point of view of a detached perspective: It is too much reminiscent of an observer-centered, pre-Copernican explanation. Here we have something that obviously is observer-centered from the point of view of this planet in this galaxy at this moment in time. So it would at first sight be a perfect match to the idea of endophysics. On the other hand, if things are looking too close to one's beliefs one should be especially skeptical.

When I was a student I thought there might be some optical distortion effects explaining the phenomenon. For example, if the universe had the structure that Einstein had originally seen for it, of a closed stationary sphere, which is an

A. Sanayei and O. E. Rössler, *Chaotic Harmony*,
Emergence, Complexity and Computation 11, DOI: 10.1007/978-3-319-06781-0_8,

alternative to the growing sphere, then light coming from the opposite pole of the sphere would be distorted optically. I hoped this distortion might explain the redshift, which did not work out. Later this skepticism to not accept something that everybody takes for granted would take on a different form. In my courses in Tübingen I had a student for many years, Dieter Fröhlich, who always asked constructively-subversive questions. So, the students acquired the habit of preferring to hear new ideas rather than to follow the usual path of thinking. We wondered jointly, in this atmosphere of mutual cooperation, whether there might be an alternative explanation to the expansionist theory that is even more natural and simple. We gave ourselves the following statements: The cosmos contains very many highly massive big objects, called galaxies and galactic clusters that are all in random motion and quite fast with velocities of about 1 percent the speed of light. Second, we have these very light particles, the photons, criss-crossing the cosmos. One can therefore make oneself a simplified picture by thinking of these two types of particles as being in mutual interaction. We asked; what would happen if these moving galaxies were replaced by moving mirrors and we would still let light pass through this collection of moving objects? We wondered whether the interaction with the motion of the mirrors might not change the energy of the photons in a systematic fashion. This is a well-known situation in statistical mechanics.

Statistical mechanics deals with systems of many particles that are all in interaction. If you come from chaos theory, you believe that this is a theory that is basically deterministic even though you can describe it in statistical terms. So one understands quite well what happens to a particle that passes through a cosmos which contains moving repulsive particles, mirrors, that reflect the photons. Although the repulsion would be quite gentle, nonetheless there would be a systematic energy change to be expected for the photons as they traverse the cosmos. It took us quite a while to realize that this picture with repulsive galaxies would not give a decrease in the energy of the photons as they pass through this billiard type situation, but rather an increase in energy. There would be a distance-proportional blue-shift, that is, the opposite of what is observed. But we had, of course, made an assumption which is the opposite of what is the case. Galaxies do not repel photons, they attract photons.

Gravitation is the main force exerted by them while moving around in the cosmos. Hence the hypothesis arose in the lecture room that, maybe, there is a systematic *decrease* in energy for photons negotiating a path between moving attracting galaxies and galactic clusters. I then gave a talk on this in December 2002 at the University of Oldenburg after having been invited by my former student, Joachim Peinke. I also had contact with Ilya Prigogine who was still active. Our statistical-mechanical picture was a multi-particle chaos type situation which was interesting to investigate. In December 2003, we published the hypothesis that this non-repulsive mechanism can explain the Hubble redshift law—that this chaos-type situation explains the decrease in energy of light-particles traversing the cosmos, in a distance-dependent fashion just as is observed. It then still took a couple of months to find an equation, published in the journal "Chaos, Solitons and Fractals" after I had talked with Roger Malina at a conference in Holland. Then I got a letter from a young scientist with whom I had published on cosmology after we had met at a conference in Switzerland, Ramis Movassagh. He had unearthed a paper by Chandresekhar of 1943 in which Chandra apparently described the same phenomenon, but not with photons going through a cloud of moving galaxies but with fast-moving stars traversing a cloud of randomly moving stars. He looked at a globular cluster of stars in which many stars are moving at random and where sometimes one particular star gets very strongly accelerated in the middle of the cluster in a so-called three-body almost-collision. Then two stars remain closely bound circling each other as a so-called "star molecule" while the large potential energy between them is carried away by the third star of the triple encounter which is then shot-out from the middle of the globular cluster at a very high velocity. Chandra wondered how these very old structures in our Milky Way galaxy still persist, when not infrequently one of their members is shot out from the dense core with a very high velocity so that the cluster should long have evaporated. These clusters in the halo of our galaxy are, as we know to date, actually the oldest structures in the universe, being more than 13 billion years old while the Big Bang scenario holds that the whole cosmos is not much older than 13 billion years itself. Chandra was able to set up a mathematical calculation using Brownian-motion type modeling (as Einstein had introduced it in 1905) showing that these very fast stars originating from the dense middle of the globular cluster will on their way out be braked at a distance by the other stars so that they cannot get lost to the cluster but are eventually slowed-down enough to remain in the cluster. So he had found an explanation for the amazing persistence of these oldest structures in the cosmos found in our home galaxy.

Sorry for interrupting: How exactly can this procedure explain the structure of the cosmos and its creation in place of the Big Bang theory?

Our idea was to find a mechanism of how light is losing energy while traversing the cosmos. We produced an example that a fast particle would be losing energy in interaction with randomly moving galaxies. The idea conveyed to me by Ramis was that there might be a connection between Chandrasekhar's result of the braking of high-mass fast particles, namely stars in globular clusters, and the hypothesis of a braking of low-mass fast particles, photons, that occurs in a cluster made up not of moving stars but of moving galaxies. Otherwise, it would be the same phenomenon.

Only quite recently did I discover that since 2007, the year our paper appeared in Chaos, Solitons and Fractals, there is an anonymous article in Wikipedia in which exactly the same comparison between Chandrasekhar dynamical friction and the idea of light getting tired when traversing the cosmos, as Fritz Zwicky had first proposed in 1929, is made. The let-down was that Wikipedia came to the conclusion that this analogy does not work. It is educating to see that such a "close encounter" of imagination is possible in science.

At this point, everybody would probably wish to hear about calculations, how strong is the effect quantitatively? The 2007 paper already contains such an estimate. The main point though is that this is just another example of Hamiltonian chaos theory. The general question is whether Hamiltonian chaos can explain, on the one hand, that repulsive fast particles statistically gain energy in many-particle interaction with more energetic heavy particles, which is the subject of a well-known discipline called statistical thermodynamics, and whether, on the other hand it can also explain the analogous situation in which the forces between the particles are attractive instead of repulsive. Is it possible that here, exactly the opposite of what is known in statistical thermodynamics occurs? Chandrasekhar had not used the model of statistical thermodynamics; he had the idea of Brownian motion in the back of his mind, while statistical thermodynamics is an applied version of chaos theory. So here chaos theory proved or seemed capable of saying a word about the cosmological redshift. There is a connection between Hamiltonian chaos theory on the one hand and the cosmos on the other. A circle harking back to ancient Greek ideas about chaos and the cosmos seemed to close itself. This was quite an attractive idea.

Would you mind explaining a little more about the structure of statistical-mechanical theories?

Thank you for this question. The chaos theory in the simplest case deals with, how shall I explain, …

Maybe I can ask this question: If you consider these mirrors and the reflection of the photons between them, this is quite like the billiard chaos. Is it possible to apply the theory of Sinai chaos here? Or is the structure of the new statistical mechanics that you mentioned different from Sinai chaos?

Thank you for this important ball thrown to me. Yakov Sinai was the first to apply chaos theory—as it was to be called later—to interacting billiards. He was very ingenious in doing that. First, he pictured a gas of several particles that are mutually repulsive in three dimensions. Then he wanted to understand what really happens. It is very hard to get a completely satisfying picture out of such a difficult mess of interactions. Maybe I'll describe his series of learned simplifications a bit: Initially he assumed that the space in question is not 3-dimensional but only 2-dimensional. Then he assumed that he had only two particles instead of many. These two particles would form two interacting repulsive round disks. But it was still not clear what these two billiards would be doing with each other, it still remained too difficult to understand. So he made an additional simplification: he nailed one of the two round disks onto the middle of the quadratic table so that only the other disk would be moving. Then he realized that he could make the width of the middle particle twice as large as before and in recompense have the moving particle shrunk to a point, without any change in the motion of the latter. It would just be the motion of a point being reflected symmetrically from a disk in the middle. Then he identified opposing sides of the quadratic table to get a linear torus, again without change of motion except for intermittent mirroring effects that make no difference.

This situation can now be replaced by—is equivalent to—a situation in which there are very many regularly spaced fixed disks, and a single point particle is moving straight across this set until hitting a disk, to get symmetrically reflected into a new direction, and so forth. My friend Harry Thomas in Basel called this ingenious picture "tennis game in an orchard": The trees would symmetrically repel the tennis ball on being hit. One could have a cut through this orchard of regularly spaced trees and there would be a point-shaped tennis ball flying between them without friction. This is the mental picture of a Sinai gas. Under this pictorial condition, Sinai could explain and prove that this system is chaotic. This was in 1970 already, 5 years before chaos would be given its name. It was one of the big breakthroughs in chaos theory.

If we return now to our context of a radically different type of gaseous behavior, possibly existing under attractive conditions, the question poses itself whether or not a similarly simple picture can be obtained if you replace the repulsive short-range force Sinai had used, with an attractive, long-range force. In this way one comes up with an analog to Sinai's problem within fixed galaxies—fixed centers of gravity—and a moving point passing across them. But this is still too difficult because it remains a multi-particle 2-dimensional problem. For quite a while we made no progress because the simplification was not yet radical enough. Then, interacting with Klaus Sonnleitner and John Kozak, we arrived at a "T-tube model" where the motion of the billiard particles is no longer in 2-D but in 1-D. In one dimension, however, two interacting particles cannot generate chaos. So this appeared hopeless before the T-tube configuration was hit upon. Image the letter T. In the horizontal bar at the top, you have one particle moving back and forth without friction, and in the vertical bar you have another particle without friction which can move only up and down. You can then make the two particles interact via a Newtonian force acting across the momentary 2-dimensional distance between them, but of which only the sliding component is effective as natural in the absence of friction. Then there are two canonical cases; you can either make the force repulsive (as Sinai had done) or attractive. Newtonian repulsion and Newtonian attraction then make for a prototype pair.

Sinai was repulsive?

Yes. The Sinai system and the T-tube are closely related. There is a moving point on a 2-D configuration space here. This is related to Sinai's 2-D real-space case. Joe Ford had told me almost 30 years ago that there exists a "theorem by Bunimovich" to this effect. Bunimovich recently returned the ball to the late Joe Ford in an E-mail. The T-tube system has the added advantage that one can safely put it into a computer because the equations are simple enough to be solved numerically with high accuracy. An unusually accurate 4th-order Hamiltonian symplectic algorithm (which roughly speaking repairs numerically caused deviations from exact energy conservation), written by Sonnleitner, allowed for the simulation of both cases (the repulsive and the attractive one, which differ only by a sign flip). When the massive particle in the stem of the T-tube was slowly moving with its higher kinetic energy and the lighter particle was fast moving, then the slow massive particle took away kinetic energy from the light fast one in a statistical fashion—just as this had been predicted for the photons in the cosmos losing energy to the slow heavy galaxies. To check on this, we of course also alternatively changed the sign of the attractive Newtonian force towards repulsion. Sonnleitner predictably observed the opposite effect. Indeed in that case the fast particle statistically gained energy showing a tendency towards energy equipartition in confirmation of statistical thermodynamics. This T-tube model proved very powerful because until then, statistical mechanics had not been based on a deterministic model, except for Sinai's gas. Sinai had looked at the equilibrium case since the moving particle's energy was constant. Here, with the two moving particles of unequal mass, we could for the first time numerically reproduce a well-known 150 years old fact from statistical thermodynamics under deterministic conditions, a tendency for energy equipartition under smooth repulsive conditions. This tendency even occurred in a maximally simplified 2-particle case of unequal masses in formally one dimension. The tendency for equipartition of energy arose as a direct consequence of chaos theory. Thus, deterministic chaos could explain the equipartition theorem of statistical thermodynamics in a formally 2-particle, one-dimensional system in a T-tube as a "prototype" system. It goes without saying that one could add more particles in more T-stems. The essential finding was that chaos theory implies far-from-equilibrium statistical thermodynamics, which was an absolutely unknown fact so far. This was the breakthrough made by Klaus Sonnleitner. It was accompanied by an even greater breakthrough because it possessed a *mirror* case—that if you invert the sign of the force to have attraction in place of repulsion, you get the opposite result. The latter was exactly what we had predicted for the cosmos. Under attractive conditions, we do not get energy equipartition as everybody would have been ready to bet, but the opposite, anti-equipartition, a new phenomenon—a Hubble-type phenomenon. A disproportioning of temperature rather than a smoothing-out of temperature. One could say: "so what?" What is the use of reducing statistical mechanics to chaos theory? The simple result shows that we have hit upon a sister discipline to statistical thermodynamics. So we have to give it a new name. It is no longer a statistical thermodynamics, it is a statistical cryodynamics, because here the temperature of the low-energy particle is decreased, it is getting colder. Cryós means cold while thermós means hot.

At the same time, new light was thrown on one of the most baffling phenomena of nature and history; we had reproduced a phenomenon that had baffled the human mind for more than a century since Boltzmann, namely, that a time-reversible system shows a time-directed behavior. We saw that we only need to invert time after a while in a simulation to have the opposite behavior in the very same system. In the one time direction, one has an increase of entropy in the case of thermodynamics, in the other, a decrease. This is well-known even if it could not hitherto be reproduced deterministically. This dichotomy got enlarged into a "double dichotomy" because after switching to attraction we now have a decrease of entropy in the forward time direction in cryodynamics. This is all very surprising because how do the particles know whether they should have the one behavior or the other if there are 4 options, two for attraction and two for repulsion? In each case when you let the system run for a while and then invert time, the opposite behavior occurs. How did it know beforehand that it had not done something in the past so as now to retrace that old path? Something like a distinction between "uncommitted" initial conditions and "pre-committed" initial conditions applies here, and this in a doubled fashion. As it turns out, the distinction between uncommitted (virgin) initial conditions and the others is well-known in statistical thermodynamics where Boltzmann first saw it in 1895 and coined the name, "hypothesis of molecular chaos." This feature means that the particle in question has no past history of motions behind it. This implies that the two classes of initial conditions have different measures, with the uncommitted one overwhelmingly more frequent. This is a most interesting mathematical phenomenon within chaos theory.

You mean, this hypothesis assumes that there isn't any correlation between velocities and they are independent of position? Why did he call this "chaos"?

He may have used the word chaos in a naïve sense that everything is out of correlation with everything else. It is also possible that he already knew about the distinction between ordinary infinity and the transfinite super-infinity which Georg Cantor had introduced, and that it is of possibly vital importance here.

Is it possible that he wanted to assume that there is no correlation between speed and position here? In this case it would have to do, not with an anticipation of chaos theory but with an anticipation of quantum mechanics?

Interesting, I believe Boltzmann only wanted to say that these particles interact in a way which does not reveal a previous interaction they had had. That it must be a totally freshly generated, new situation—that they do not know each other beforehand. This may have been enough to let him use the unfamiliar word "chaos." We need to find a specialized historian of science to really clarify the roots of this notion of chaos introduced by Boltzmann.

But scientifically, you do not see any relationship between his thesis and the definition of chaos?

There must be a deep connection here because this is a chaotic system, as we know. If we choose a non-selected initial condition—I mean one at which the system has not arrived by having been let run for a while, but an arbitrarily chosen "random" initial condition in a Cantor continuum—then we predictably always find in the corresponding simulation that the big particle will give away energy to the light particle when the force is repulsive (and take-away energy when it is attractive). However, this idealistic mathematical sensitivity can now be studied empirically in a finite-accuracy setting for the first time. Infinity can be finitely simulated, so to speak. When we stick to the repulsive case—the only one known up until now in statistical mechanics—then if the two particles had not been interacting before in the simulation, they invariably must give you this time-directed behavior—of the light fast particle getting more energy from the slow heavy particle, even though the time direction is not specified, that is, in either direction of time. So the system shows the same coordinated behavior in both directions of time when you start out from a non-selected initial condition, which is what Boltzmann had seen in his mind. But now the same thing can be reproduced in a 2-particle deterministic situation of finite accuracy, which is a very surprising fact. As a bonus (and still more spectacular) the opposite behavior occurs under attractive conditions. So indeed the hypothesis that a light-in-mass particle loses energy when traversing a cosmos of randomly moving heavy-in-mass particles called galaxies or galactic clusters appears to have been verified numerically in this very simple 2-particle T-tube system by Klaus Sonnleitner in his dissertation of 2010.

Do you see any relationship between this idea and the theory of the so-called "Poincaré recurrence"? The latter period is assumed to be so extremely long, practically infinite, so that it does not affect anything in science. On the other hand, it is true that our present situation is valid in finite time. Do you see any relationship whatsoever between the Poincaré recurrence and the phenomenon you mentioned?

By all means, this chaotic system is bound to obey Poincaré's recurrence law. Boltzmann also knew about this recurrence. At first it was presented to him by his Viennese colleague Loschmidt as a counterargument to his time's arrow which he eventually accepted. He realized, however, that the recurrence times are very, very long. So the increase of entropy will eventually have to be followed by a decrease of entropy again, but the recurrence time is unfathomably long mathematically. They must remain very long in the presence of a finite numerical accuracy; this is a new question which can now be investigated. We have thus this 2-particle chaotic system in which we can really answer the questions brought up by Poincaré and Boltzmann more than a century ago. However, since Sonnleitner found a numerical "exit point" eventually at which the motion became dull—periodic—for reasons of numerical inaccuracy, the problem will not prove as easy to investigate as it appears at the first moment. It will no doubt, be very much fun to re-open this whole issue from the point of view of this trivial, completely transparent chaotic 2-particle system. I can add here that the system most likely can be simplified still further. We can make the particle in the stem of the T "infinitely heavy" in a stepwise manner while making the interaction proportionally weaker so that the effect on the light horizontal particle is unchanged. The only difference would then be that the fast light-weight particle no longer acts back on the slow heavy particle—as we may assume also holds true to a very good approximation for the back-action of a passing photon on a moving galaxy. Then the system becomes mathematically even simpler. It actually would just be a periodically forced 2-variable Hamiltonian oscillator. The light horizontal particle would then be periodically forced by the vertical particle, and the Poincaré map would be as simple as that of any other periodically forced Hamiltonian oscillator. It would be a periodically forced single-degree-of-freedom system—the simplest type of Hamiltonian chaos conceivable. While all the "comfort" of the T-tube paradigm would survive, all of Poincaré's (and the later developed) instruments to numerically study chaos in a periodically forced oscillator could be applied to this system. But, as stated, this is merely a hope at the present moment in time because we have not checked this possibility numerically for lack of funding. On the other hand, to a mathematical mind this simplification is not actually needed because the two particles have a 2-dimensional (if non-planar) Poincaré cross section in the 4-dimensional state space since the motions occur on a three-manifold. Everything would be facilitated if the idealization I described works. As stated, I predict that the heavier particle need not sense the motion of the fast particle just as the galaxies surely do not care about individual photons passing through. The phenomena are predictably insensitive to the ratio between the masses approaching infinity. Thus indeed the simplest and oldest type of chaos theory, invented by Poincaré in the late 1880s, can now be brought to bear on cosmology.

The question is, now that you explained the structure of the simple model of two classical masses, how do you solve the problem that photons are rest-mass-free particles?

This is a bit tricky. One could simply say that everything that can be predicted about photons traversing the cosmos will hold true also for cosmic ray particles traversing the cosmos, that is, for near-luminal protons. The fast protons would be a mechanical analog to the photons and we would still understand what happens. The protons would still fly with almost the speed of light, so that formulas from special relativity need to be recruited in the final analysis. But the classical theory would not need quantum mechanics to be valid for once—quantum mechanics would follow suit. Of course a full-fledged relativistic-and-quantum mechanical treatment will need to stand at the end.

Then the last question in this session has to be: Can we consider the relativistic special case of the T-tube? Suppose a moving observer wants to see the structure; is there something different, something interesting or new, if we add the velocity effects of relativity theory?

It is amazing: Just as quantum mechanics is not really relevant here, so also general relativity is not really relevant here. The main phenomena are classical. There are slight modifications when we add the fact that space time is no longer Euclidian according to Einstein, but this would not have an essential effect in the classical-chaos based paradigm which has been found to apply here; where chaos is a third leg between quantum mechanics and relativity.

When we carry out this experiment in two different relativistic frames, do we see any new or interesting things?

What we find to hold true at low speeds classically, predictably remains valid if relativistic speeds are allowed-in. Qualitatively speaking, this would be a perturbation that makes no difference. The ostensibly outdated classical thinking of Clausius, Poincaré and Boltzmann would be sufficient to explain the main phenomena. Both quantum mechanics and special relativity would smoothly join in without changing the qualitative picture. But then we also know from experience that such are the best occasions for breaking new ground again. This is virgin territory.

Chapter 9
Relativity Theory, Speed of Light and a New Theorem

We talked on cosmology in the previous session and I suggest we continue a little more. I would like to ask you, what are your ideas about the speed of light, about relativity theory, and what are your new contributions to these topics? Which insights do you consider important and what are the implications regarding cosmology?

Cosmology is a very exciting field and it is also a field which is very safe to study because it has no direct implications for human life except if one detects a meteorite bent on a collision course with planet earth in observational astronomy. Cosmology is intimately related to Einstein's thinking because the speed of light is so important for the measurement of large scales and c, the pertinent constant, also is a deep mystery. No one can explain why it has a finite value and why the latter should be a universal constant. Einstein in 1905 took c to be a universal constant as it still remains in special relativity. Two years later, however, he reluctantly gave up on the idea of the universally valid constant c, after he had hit on the possibility of understanding the problem of gravity with the aid of special relativity. He replaced a tower on earth by a long-stretched rocket-ship in constant acceleration. In this fictitious scenario, he saw that when light is ascending from the bottom of the rocket-ship to the tip while the rocket-ship is constantly accelerating, then during the traveling time which it takes the light to reach the tip from the rear, the rocketship picked up further speed. So when the light arrives upstairs at the tip, the tip always possesses a constant speed relative to the light's source at the bottom. So there is a predictable Doppler effect here—a redshift—compared to the rear end of the ship.

This was a "hard" inescapable prediction regarding gravity as well. It represented his next deep original insight 2 years after he had discovered special relativity in close parallel with Poincaré (but with two farther-reaching predictions). The new mental picture of the extended long rocketship in constant acceleration is the source of the eventually arrived-at theory of general relativity. There existed a definitive chance to fully understand space and time in gravitation based on the strange feature of the speed of light possessing a constant finite value. But then a

A. Sanayei and O. E. Rössler, *Chaotic Harmony*,
Emergence, Complexity and Computation 11, DOI: 10.1007/978-3-319-06781-0_9,

minor catastrophe occurred. In that same *article* of 1907, Einstein realized that the speed of light can no longer be a universally constant speed. For if you look down, from the tip of the rocket ship onto a lower level where momentarily a light ray is passing horizontally (in a glass tube filled with thin smoke to make a passing light pulse visible), predictably the speed of light down there is reduced. For example, if the accelerating rocketship is assumed long enough that the clock speed is halved, also only half the speed of light is seen at work down there compared to upstairs. Thus, the universal constancy of the speed of light c as a globally valid law of nature was suddenly shattered. It only could be said that locally, the speed of light remains constant. If you look at it from the tip, you see that it slowed down in the horizontal direction. This conclusion deeply worried Einstein. For almost four years he stopped publishing on the topic of gravitation because of this drawback. Eventually, his close friend Paul Ehrenfest, who visited Einstein in Prague where Einstein had obtained a professorship, could bring him back to thinking about such questions again.

Ehrenfest brought up the topic of a rotating disk and the types of transformation that apply there. In the attempt to help Ehrenfest with this, Einstein inevitably harked back to the question of gravity—and would eventually succeed in formulating the theory of general relativity. Hereby the distressing finding that the speed of light is only locally but no longer globally constant remained a feature of the emerging full theory.

Up to this day, the topic of the constancy of the speed of light remains a contended issue. In Tübingen, the "chaos school" kept the problem on a low burner for decades. It can be described exactly by mappings—a familiar technique in chaos theory. We looked at what happens if you send a light pulse up from a lower level in the rocketship and reflect it back down again. You can compare the intervals generated in the zigzag between upstairs and downstairs. You can find, for example, that when you have a one second interval upstairs between emission and re-arrival, you may have only half a second downstairs. Two light rays of one second distance sent down and then up again have the original one second distance again. However, downstairs the interval between the same pulses is consistently half a local second. Thus, there exists a 1–1 mapping between the slower time intervals downstairs, half a second, and the faster time intervals upstairs, one second. This is quite surprising. Ordinarily such mappings, when you have expansion in one of two legs, do generate chaos. Hence we expected to obtain chaotic behavior if one produced iterates between sending light upstairs and downstairs. This seemed to us like a promising new way of looking at Einstein's ideas—but to our surprise there was no chaos. The whole recurrence, as such maps are called, proved to be bijective across the two differing time scales. Thus we were forced to think twice about the deep paradigm of the extended long rocketship in outer space which Einstein had invented as a Sci-Fi type riddle in 1907. This amazing fictitious situation of the totally transparent extended rocketship is usually called the "equivalence principle" because it postulates that there is equivalence between an extended accelerating rocketship in outer space on the one hand, and an equally long tower standing on earth in gravity on the other. This idea—to consider in special relativity acceleration at two different height levels simultaneously—was Einstein's deepest insight in my eyes. He always called it "the happiest thought of my life." For it was so non-trivial to arrive at this trivial-appearing scenario. People know quite well today how he hit on this insight. I do not know whether I should tell this little story from Einstein's life in 1907.

He had a heavy stomach when he entered his office—or rather the common office—in the Swiss Bureau of Mental Property. The big window stood wide open and he fleetingly wondered what would happen if he jumped right out of the window: would his stomach feel lighter? He realized in a flash that "of course" it would feel lighter because when I am falling, the stomach by definition cannot pull down on me since the whole body falls freely. So gravity would no longer exist when I am in free fall. To date everyone has seen TV reports of astronauts living without gravity in a satellite or of Stephen Hawking floating with his wheelchair weightlessly in a parabolic flight. So it is true that one indeed can be weightless. The flash-like intuition into weightlessness enabled the 28 years old Einstein to realize that he could get rid of the phenomenon of gravity altogether—by falling. When falling, there is no gravity left, one is effectively in free fall as in outer space. Hence there is a way to magically switch-off gravity. Because of this foreign insight—by the way also arrived at by Newton as Thibault Damour dug out—Einstein knew for sure that he would be able to formally completely unravel the essence of gravity. His asset over Newton was that special relativity with its universal speed of light c happened to form a universal instrument in his hands that Newton lacked. After this excursion into the history, we can at last come back to the mentioned "Einstein effect" that light seems to be "creeping" downstairs in gravity.

Do you believe that the speed of light is constant? Did Einstein make a mistake in that context, or did he not?

He founded special relativity on the universal constancy of c, but 2 years later felt he had to sacrifice this feature. The story is maximally exciting in my eyes. We know the speed of light is constant in special relativity and hence also in Einstein's rocketship. Then, if we look down in the latter from above, c appears slowed-down transversally downstairs. The question is; can we reconcile this fact of 1907 with the global constancy of c of 1905? I had the good fortune of having very originally minded students over many years. Dieter Fröhlich and Heinrich Kuypers helped me, and together we found that the situation is a little bit different than we all thought. The point is that when the clocks are slowed-down downstairs, this is not the only phenomenon that occurs down there. Quite recently a paper titled "Olemach" was devoted to unraveling this. O, L, M and Ch were combined with two arbitrary vowels to make the theorem more pronounceable. O means omega, the rotation rate of a given frictionless wheel downstairs in gravity. L means length, M means mass and Ch means charge. We can haul the freely rotating wheel (a bicycle wheel suspended at its hub, say) up and down. The rotation rate is then reduced downstairs because a rotating wheel is by definition also a clock. There is a theorem in mathematical physics, the Emmy-Noether theorem for angular momentum, which guarantees angular-momentum conservation in the presence of a rotational symmetry. Since rotational symmetry is preserved in Einstein's extended long rocketship, this theorem automatically applies. It demands that angular momentum *L* is constant. But *L* is equal to rotation rate *omega*, times mass *m*, times radius *r* squared, for a rotating wheel.

r, the radius, squared times mass m times rotation rate omega?

Yes, one can almost remember this rule because it is intuitive. We know that this law is universally valid in the absence of friction. Thus it must also be valid at the bottom of our accelerating Einstein rocketship. If the rotation rate omega is reduced at the bottom of the ship, owing to the Einstein time slowdown, then either *m*, the mass of our rotating frictionless wheel or *r*, the radius of our frictionless rotating bicycle wheel suspended at its hub in a frictionless manner, must be changed, or both. Since the product of *omega* times m times *r* squared is constant in the absence of friction, the change in *omega* that has occurred "must" be compensated-for by at least one other change. There are several hypothetical laws that could be conjured up to apply here. But we are decisively helped by quantum mechanics which tells us that mass *m*, which makes up a part of this omr^2 expression, is reduced downstairs in proportion to the slowdown of time (or equivalently omega) because local photons have less mass-energy down there. They have a lower frequency according to Einstein as we saw. Photons are locally inter-transformable with particles, into particle masses, as is well known. Positronium annihilation (and creation) is a case in point with which everyone is familiar today since it is used in positronium-emission-tomography. Everyone knows about the "PET scan" today. A PET scan, in which a positronium atom is self-annihilating into two photons of prescribed energy, works equally well at the bottom of a tower (or the extended long rocketship) as at its top. This holds true despite the fact that the frequency and hence mass-energy of the photons emitted down there is reduced by the Einstein redshift factor. Recall that the clock slowdown means that photons have a slower frequency downstairs (are redshifted). Hence all masses locally at rest downstairs must be reduced by the redshift factor along with the positronium atom. That is, we know for sure now that in our product (omega, times m, times r-squared), *omega* and *m* are reduced in parallel. So, only *r* remains. Since the law says that the whole thing omr^2 remains constant, with *o* and *m* gone down in parallel, *r* must have gone up by the same factor. Only then is angular momentum (omega times *m* times *r*-squared) the same as before, when the wheel is lowered reversibly to get slower and then pick up speed when hauled up again. So we see that size (*space*) changes along with rotation rate (*time*).

How new is this result?

The size change in gravity has been seen by quite a few people, including Ron Hatch and more recently Richard J. Cook, a physics professor at Colorado Springs, in a paper published in 2009 on the arxiv. Thus one indeed has a new fact in Einstein's theory. It changes virtually everything also in cosmology by the way—it is embarrassing. If indeed along with m, also L is changed downstairs, we have both a slowdown in time and a proportional increase in size. Therefore, the ratio between length and time is constant. So it suddenly turns out that the speed of light c is a global constant after all. You then no longer have to assume that the poor people downstairs have to live with a smaller speed of light without noticing. It now follows that it is no longer possible to assume that anywhere in physics—for example, in general relativity—the speed of light were not globally constant.

Also in cosmology?

The result extends to cosmology. If one believes in a cosmos that is expanding, one has to assume that the speed of light is only locally constant. For only then can the whole of space expand. This is the case in the Big Bang model. So the Big Bang model is, for a very fundamental reason ruled out by the Olemach theorem. In this way, playing around in your mind with a rocketship that is accelerating in outer space allows you to come up with very surprising implications. Einstein is still alive with his way of thinking. The clock slowdown discovered by him by sheer thinking has corollaries. The Olemach theorem is only a completed version of Einstein's happiest thought. One is then, of course, forced to ask the follow-up question: How about general relativity? How can the speed of light c, which is *not* a global constant in general relativity, suddenly become a constant in general relativity? Fortunately, this does not lead to a contradiction. It turns out that in the Schwarzschild metric, which is the most important solution in general relativity, there indeed exists an implied observable which represents the correct measure of length using which, the speed of light is globally constant. This fact we actually found before discovering the same result also in the constantly accelerating rocket ship. This earlier finding we called the *gothic-R* theorem. The "gothic R" is not the same observable as the familiar capital R, nor the little r that are both well known in the Schwarzschild metric. The *gothic R* is written in a more convoluted form than the capital R. For this mathematically admissible new length variable, the speed of light is globally constant. The Olemach theorem, which remains applicable in the Schwarzschild metric, only confirms the gothic-R theorem on a less "noble" level as it were. A very conspicuous consequence of the new state of affairs valid in general relativity in the wake of both theorems is that metrology is strongly altered. Many numerical measures in the physical universe are changed. This is because if c is globally constant, then several other formerly global constants become merely locally valid in exchange.

It appears that a whole basic field in physics, cosmology, is markedly affected by a little new implication of Einstein's happiest thought. One of the new effects is that cosmology can no longer scientifically assume an expanding model of the cosmos. It is perhaps interesting to add in this context that another fundamental theory of physics, statistical mechanics, cuts the same way. A new second branch of statistical mechanics implies that light is losing energy while negotiating its way through randomly moving galaxies. This Zwicky-type result converges with the Einstein type result just discussed. So it is both, a new metrology and a new cosmology which come out of a re-thinking of what really happens downstairs in an accelerating extended rocketship—the "Einstein couch" (as it can be called with a glance to Freud). There is one prominent specialist here—Wolfgang Rindler—who devoted his life to thinking deeply about the accelerating rocketship and wrote the most influential book in the field ("Relativity, Special, General and Cosmological") who does not dissent so far with the sketched new results. But I also have to admit that the number of specialists in general relativity who so far signaled concurrence with these results is countable on one hand. It is an interesting experience to see how intrinsically sluggish communication is in science—but this is another topic. It is a well-known fact that after finding something new, one is confronted with all kinds of surprises and that, the more surprises there are, the more sluggish the appreciation is in coming. But the silence about *c-global* and *cryodynamics* could of course also mean that both results are nothing but two mirages seen by the same camel train.

If you stick to your result, is it not an automatic consequence that an alternative to the Big Bang is needed? Can we have another fundamental explanation of cosmology—after the Big Bang model?

I believe that I can show that a new explanation of the main phenomena in cosmology is necessary in the wake of *c-global and cryodynamics, Inc.* The most fundamental cosmological phenomenon is the Hubble law, for which already several Nobel prizes were awarded and which is an absolutely indubitable phenomenological law. Hubble himself did not get the prize but 2 years ago, Perlmutter, Schmidt and Riess received a Nobel Prize for a modification of the Hubble law at high redshifts that implies a belated honor for Hubble as well. They found that the Hubble line is no longer a straight line but possesses a little hook at the end. It was for this little hook that the prize was awarded. But this very hook now needs a new explanation. If the Hubble line is not caused by expansion, then the little hook also needs a new explanation. The deviation from linearity can be explained easily by recalling that a fractal cosmos has big holes, the so-called voids which occur in larger and larger versions the farther you go out in distance. If you take this fact into account in an expansion-free context, it turns out that this little hook can be explained along with the redshift itself by the discussed Zwicky-type dynamical friction model. So there exists a new synthesis which seems to cover everything that is known in cosmology so far. Nevertheless it is also important to look at all the different independent confirmations of the Big Bang that are part of the "modern synthesis"—with the so-called "cosmic microwave background radiation" (MBR) taking first place after the Big Bang itself.

Another question: If we suppose the result which you mentioned, Olemach, is true…

One never knows whether one does not make a major blunder in one's mind—

…then, what is the consequence if we apply this result to the Einstein equation?

Thank you for harking back to the fundamentals. The best-known solution to the Einstein equation is the earliest one, found by Einstein's friend Karl Schwarzschild a few months before the Einstein equation got finalized. This "Schwarzschild solution" happens to contain the constant-*c* result already, in form of the gothic-*R* observable. Einstein was so meticulous in his craftsmanship that even though he believed *c* not to be globally constant, his equations are compatible with a globally constant *c*. It is only certain particular solutions that cease to be valid—like the expanding universe solutions of Friedmann and Lemaitre. Note that even though the Einstein equation itself is not changed by the globally-constant-*c* result, its interpretation will need to be changed in many places.

So, we can go back to the Einstein equation and simplify its understanding and find new solutions? I mean, your results allow one to ignore some assumed implications like the expansion of the universe and presumably others?

Right, the expansion model represents a special solution to the Einstein equation. This solution ceases to be physical, and there are other solutions that cease to be physical—like the accepted combination with Maxwell's equation, the so-called Einstein-Maxwell equation. It is not Einstein's fault that this hybrid equation is no longer correct since the gluing-together point belongs to an outdated interpretation and hence the combination needs to be re-constructed. There is also a famous solution to the Einstein equation alone, the Kerr-Newman solution, which like the expanding solutions ceases to be valid. It described the properties of space and time around rotating black holes. In return one obtains new rewards. Here it is the existence of a so-called Reeb foliation around any rotating black hole. So we have an on-going revolution in general relativity, even though the Einstein equation itself is not affected. It is only several derived solutions that prove to be unphysical owing to an error of interpretation.

You published a paper before Olemach which you called Telemach. What is the difference? Do we have the same result in two versions, or are they completely different?

I believe they are very different. Telemach means that *T, L, M* and *Ch* are changing by the same factor either in the numerator (twice) or in the denominator. *T* (time) in Telemach is replaced by *omega* (rotation rate) in Olemach. In this sense, the two theorems are very close. But the explanation of the two theorems is quite different. Telemach is much more complicated to derive. Olemach was found later and is more elegant (I think) because it is based on Emmy Noether's rotation-symmetry theorem. Olemach helps one better to enter the exciting field of global-c general relativity (cGR).

I think the Ch-part in both theorems may prove to be decisive in the end—the prediction that charge changes in a gravitational field. Maybe you should put it up as the main new result?

I see. We started out with mass changing downstairs near the bottom of an extended rocket-ship in permanent acceleration. Once the local rest mass has been found to be changed, one is obliged to ask the question; what about charge? An electron has both rest-mass and charge. We know that the electron's mass is reduced downstairs along with photon mass energy—the example was positronium creation-and-annihilation. We also know that locally, the laws of nature must be unchanged in any locally momentarily co-moving inertial system. That is, when a particle is momentarily released into free fall, it "in general" does not change its properties compared to its still fixed neighbor. So the ratio between mass and charge remains unchanged locally. Therefore, *Ch* or q (for charge) must be changed in proportion with rest mass m. However, the "law of charge conservation in physics" is as old as the American democracy because Benjamin Franklin first formulated it. It represents one of the fundamental pillars in physics. Now this venerable law is suddenly gone if Einstein's thinking about the accelerating long rocket ship was sound. How can this unavoidable "revolution" be made plausible? There exists a proposal made by James Clerk Maxwell which I once found in Martin Goldman's inspired book "The Demon in the Aether": that there exists a golden road, a trick towards coming up with important new findings. Go to the beginning of a "gold vein" where someone by happenstance—Maxwell was very modest in his attitude and a believing socialist—has found something very important in physics. Then go back to the first beginning of such a gold vein—and dig in the neighborhood of that point. Then you will in general find a new gold vein. He even gave this option as the only reason why every worker in science is obliged to publish: Not to gain credit from one's contemporaries, but only to contribute to the thick layer of new sheets that like leaves in autumn cover the ground so that later the beginning of a gold vein can be retrieved. For one beginning gold vein has always other gold veins taking off in its neighborhood. This advice is still golden, it appears. I would say the equivalence principle represents such a miraculous place, a bounty.

I'll go back a little; you found that charge can change after all. What then about the law of conservation of charge?

Conservation of charge is gone. This is a major catastrophe.

If you say so. Do we need a new electrodynamics theory, even in the classical domain?

Only, if you combine electrodynamics with gravitation. The combinations that were accepted so far have to be re-worked because they were shown to be false. So it is really a kind of revolution in physics that is triggered.

If we consider that charge is not conserved any more, how do you apply it in studying black holes? What are the interesting implications regarding black holes?

Black holes are the most interesting implication of general relativity and the most impressive one at that. If the speed of light is globally constant and we already know for sure that time takes an infinitely long time to reach the horizon of a black hole from the outside (and again back), then the distance from the outside world to the black hole automatically is infinite in contradistinction to what is textbook knowledge. This is very hard to understand intuitively because you can fly around a black hole at some distance. You can put a sphere around it which has a finite diameter and a finite surface area. On the other hand this very same sphere encloses an infinite volume of space if it is true that *c* is globally constant. So the properties of black holes are totally changed. Oppenheimer and his student Hartland Snyder, a mathematics-educated former truck driver, discovered the black hole as a physical solution in 1939. In their paper they already showed that an external observer can if released into free fall reach the horizon of a stellar black hole within 2 days (and can also in principle come up again in another 2 days according to his wristwatch if there was a strong enough trampoline mounted on the horizon to shoot him back up). But this wristwatch time is not the outside time. Oppenheimer and Snyder also saw that during this travelling down or coming up, in either case an infinite amount of outside time is passing. Usually, people think that the two time scales, the external one and the falling astronaut's wristwatch time, have equal rights. This is what democratic thinking suggests. Since the astronaut can fall-in in just 2 days, the horizon should be only two light-days away. This is not true because even light takes an infinite amount of outside time to get down or come up from there. It goes without saying that an astronaut cannot possibly outperform a light ray. With the new constancy of the speed of light now, suddenly all solutions acquire a mutually consistent status. That is, we indeed have an infinite distance and volume inside a finite sphere. It is the topological opposite to what my friend Benoit Mandelbrot found—that you can have a finite area or volume inside an infinitely long periphery. If you idealize a coastline so that even atoms can have infinitely finely crafted surface properties, then the coastline of England becomes a real fractal of never stopping self-similarity and infinite length, even though comprising a finite area. Here, we have just the opposite: a finite circumference and an infinite volume. This is not really surprising, or is it? In differential geometry, there exists something called the Newtonian…

Pseudosphere?

Exactly, the pseudosphere has the same diameter and the same circumference at its largest diameter as the corresponding sphere, and the same volume as the latter, and the same surface area, but is not isotropic on its surface. Rather than having a positive curvature, it has (the same) negative curvature everywhere. It is a very surprising object in geometry. So, differential geometry can easily deal with the strange fact that a finite circumference can surround an infinite volume. In the case of the black hole, the space around it has the shape of a "generic 3-pseudosphere." If we omit one dimension, as feasible, it looks like two infinitely long trumpets that are glued together at the rims of their outer mouths. The trumpet has a finite maximal transverse circumference at the rim. If you are an ant that has successfully walked around the flat rim, you think it cannot be very far to reach the middle of the plane that you just circled. However, in this orthogonal direction, the trumpet happens to be infinitely long. So it will take the poor ant an infinite time to reach the middle of the trumpet. All of this is easy to understand geometrically. Only for some reason, the described change of size implicit in the longitudinal direction got overlooked as a generic implication of general relativity for 98 years.

Suppose we want to produce black holes in a lab on earth. Why do you believe this is dangerous based on these results?

We do not know for sure that it is dangerous, but we do know that these black holes have completely new properties compared to what everybody thinks so far. For example, they are no longer charged. They cannot be charged because every in-falling charge will when approaching the horizon become infinitely weak along with the in-falling body's rest-mass compared to the outside. The result implicit in Olemach and Telemach, that black holes are not charged, means for example that they cannot be detected in a high-tech detector, if one tried to produce them on earth in a machine, a so-called collider. For their charge is the only feature by which they can be detected apart from Hawking radiation. The latter—which fact I did not mention so far—is also gone, owing to the infinite distance of the horizon from the outside world—the infinite length of the trumpet's neck and mouth piece. So, black holes have radically new properties. Also, they are never finished in finite outside time as one always knew with Oppenheimer and Snyder, but for some reason forgot to remember. One did not think carefully enough about the simplest facts described in their paper of 1939. I must therefore admit that I was unhappy with the prospect of black holes being generated down on earth. I advocated the moon instead. At the present moment in time, we enjoy a break in the experiment for 2 years. I hope that it will not be resumed until one could make sure that indeed no danger is involved.

I have two questions here, maybe three. The first is: What do you think about the famous Hawking radiation?

Hawking radiation is unfortunately non-existent. This follows from the infinite distance to the horizon. Hawking radiation deals with a local quantum fluctuation, a pair of virtual particles, say, an electron and a positron, that arises out of nothing for a moment in order to jointly disappear again in a short enough time that the uncertainty principle is not violated. That is what Hawking assumed. But Hawking in addition also assumed that the horizon is quite close. So if there is this pair of particles, one of the two could disappear behind the horizon in a maximally short time—before the time was up when the second had to come back to its cousin to reunite with it and disappear. If it was true that in the proper time of the loitering first virtual particle, the second could reach and penetrate the horizon to be prevented from coming back—as Hawking assumed—, then Hawking radiation would indeed exist. Now, in the proper time of the waiting outside partner, the adventurous one will need an infinitely long time to reach the horizon and disappear behind it. Therefore the waiting partner will not be orphaned because the twin is bound to return in finite time. It can return in due time to re-unite with its virtual twin particle since it could not manage to disappear for good within the proper time of the partner as Hawking assumed. Hence there is a little error in Hawking's assumptions owed to the fact that he grossly underestimated the distance to the horizon of the black hole from the outer universe. This error is very forgivable in view of the fact that Oppenheimer and Snyder fell into oblivion. However, it unfortunately has this big implication that Hawking radiation is non-existent so that black holes can no longer evaporate. I hope he is not angry with me if I say so.

You mentioned the two words, "fractal" and "cosmos." What is the meaning of a fractal cosmos?

I had a long discussion about this with Benoit Mandelbrot. He casually remarked that astrophysicists-plus-cosmo physicists are the most problematic subgroup of scientists he ever had to deal with because they are so dogmatic. He had felt the sting of their hostility when he came up with a very simple idea. The famous "Olbers paradox" which was already known to Kepler was given a new explanation by him. It represents a miracle at first sight: Why is the night sky dark? For if we have stars at all distances as astronomers thought originally in a Euclidean universe, then even though the stars get smaller in their visible area with distance, their number gets up by the same square factor so that the night sky ought to be as bright everywhere as is the disk of the sun. The Olbers paradox was solved by Hubble, as everyone learns, when he discovered the redshift law—that the light from distant stars and galaxies is red-shifted and therefore is no longer as bright as the light from more close-by galaxies. But this is not the only reason. In 1907, a precursor of Mandelbrot, Chevalier Fournier d'Albe in France, discovered the fractal distribution of galaxies—which had not yet been fully identified as being what they are in confirmation of Immanuel Kant's conjecture of 1754. The proof of the conjecture—the identification of Cepheid periodic stars in the Andromeda galaxy—was only given in 1925 by Hubble. Fournier in 1907 had mapped out the first picture of the fractal distribution of galaxies. The fractal distribution of galaxies also suffices to explain Olbers paradox as Mandelbrot stressed which caused a counter reaction from the astrophysical community.

What is the meaning of this fractal distribution?

Fractal distribution means that you have a structure like a sponge. You have holes of all sizes. The whole structure looks very much like the Menger's sponge. Or take Swiss cheese. Therefore, everybody is intuitively familiar with fractals.

Do you mean some self-similar things?

Precisely, we have such a self-similar structure in the distribution of galaxies. There are voids of all sizes.

Do we also have a fractal dimension?

Yes, the fractal dimension has been measured for the cosmos, in many independent studies. It turned out to be less than 1.2, as I remember. If one makes more comprehensive measurements at all wavelengths, it is possible that the fractal dimensionality of the cosmos will become still smaller and even approach unity.

My last question for this session: What do you think about some famous constants in cosmology? I am referring to a paper published around 1937 by Dirac concerning some constants in cosmology, with Einstein apparenty not agreeing to this idea. What is your opinion?

Dirac assumed the standard expanding cosmology accepted at the time. Then you have a finite duration of the cosmos, a finite volume, and a finite mass. Under these assumptions his realization that we have the strange number of 10-to-the-40 here, is absolutely correct. It only breaks down if we suddenly have an unbounded fractal cosmos. It could also be an infinitely old cosmos, and we might have a recycling mechanism. It is only in that light that I disagree with Dirac's hope, not with his arguments. There happen to be very interesting exciting new findings here, supporting that the cosmos may be eternal and infinitely old at the same time, whereby one needs a recycling mechanism. This option, favored by Mandelbrot, was seen for the first time around 500 B.C. by Heraclitus who had a succinct two-word description for the cosmos: "metabállon anapaúetai," which means "metabolizing it rests," that is, remains essentially unchanged. So it is both metabolizing and persisting, recycling. That is exactly what we would need if it is true that Einstein's equation supports an infinite fractal cosmos—as it may actually do.

Chapter 10
Chemical Evolution, Travelling Salesman Problem

Let us discuss a new topic. In Chap. 9 *we talked about cosmology and the Olemach theorem. Now I would like to propose a combined question; can one think about evolution from a dynamical-systems point of view?*

Very early on, I got intrigued by the phenomenon of evolution. I always looked at it as a physical problem. When I studied for my second M.B., the medical bachelor, I came across a little book which referred to Ludwig von Bertalanffy. As mentioned, Bertalanffy had coined the notion "fliessgleichgewicht" (flowing-equilibrium) or steady state as a dynamical phenomenon, exemplified by a flame which is changing all the time but nevertheless looks unchanging. I learned about a thought experiment that Bertalanffy had made; if you put a candle into a closed box with holes in it (the "black box" had just been invented in cybernetics), and you put a mouse into another such box with holes in it, and you measure input and output, then the two boxes will be indistinguishable in terms of chemical input and output. When you do not include listening, for example, you might think there are the same things inside the two boxes, because oxygen is consumed and carbon dioxide is generated (as we had touched upon). So from a chemical point of view, a flame and a mouse are quite similar according to Bertalanffy. This triggered something in my mind. A flame is not alive, a mouse is alive: what is the difference between the two after you saw the similarities? For some reason, a third element of comparison came to my mind, namely, rust. I mean the crust on the surface of iron, the oxygenation of iron. I remembered from high school that rust has the property of being autocatalytic; and fire, of course, is autocatalytic too, as everyone knows. When fire is spreading, everything is being eaten in its reach. Life has the same feature of trying to spread and to self-multiply by autocatalysis. So autocatalysis is the common denominator and the whole thing is chemical. If you believe in reductionism in modern science, then you would say that life is chemistry in its essence. So the question arose: What is the difference between fire and life? Konrad Lorenz would later tell me in this context that Gottfried Herder had asked the question; "What was lacking to the most human-like animal that it did not become a human being?"—meaning that this type of question is a most powerful

A. Sanayei and O. E. Rössler, *Chaotic Harmony*,
Emergence, Complexity and Computation 11, DOI: 10.1007/978-3-319-06781-0_10,

way to arrive at new answers. I could suddenly see in my mind that if you had a kind of rust that cannot just produce more rust, but if the chemical reaction were more complicated, then after a while as a byproduct of more and more rust being produced, some new substance may be produced as well. This new substance might, under certain conditions, also produce more and more of its own kind. So suddenly there was an idea of chemical evolution in my mind. The conviction arose that this model, that this comparison between flame and life, enabled an explanation of the thing in question, the whole thing, life. For when you have a chemical system that is autocatalytic and at the same time can produce modified or differing substances, then this represents a highly unstable scheme. Intuitively it was clear to me that this suffices to predict that life is bound to be generated in a chemical system that has enough degrees of freedom to produce autocatalysis and to produce different substances, some capable of joining into the autocatalysis so that a shifting improvement would set in automatically. This was very convincing to me at the time, although it would take me about 8 years to write a long manuscript in which this was described mathematically, and 10 years to publish a first short paper, first with differential equations and later as a chemical growing automaton à la Paul Erdös.

Evolution depends on the difference between zero and nonzero?

To be or not to be, yes. A way to look at the same thing graphically goes like this; on the left-hand side, one has a very large set of chemical substances in the form of a vertical column of small circles (empty dots) representing chemical substances. For example, carbon which can produce long chains with side groups, and water (on Jupiter instead of the carbon we would need borane and nitrogen which, taken together can also produce arbitrarily large chains, of borane rather than carbon atoms). Water would also be such a dot, a filled-in one (black dot) among very many dots, almost all of them not filled in initially. Only some would have a nonzero concentration, for having a constant influx so not to get empty and remain black. Photons would also be among the latter. Time goes to the right in the scheme. Then this "invisible machine" (since almost all dots are empty) would be left to itself, along with its huge set of pre-formed possible reactions, all represented by transparent—initially not filled-in—Teflon-like tubes. It would be a virtually infinite network of possible reactions, whereby the substances on the left in the scheme could be freely moved or copied and pasted horizontally to the right to as a function of time, let them partake in a network of tubes and knots (dots). Almost all tubes and almost all glass balls (knots) would be empty at first. But time would go on. This almost everywhere invisible (not filled-in) machine indeed exists in the sense of mathematics, even though not as a physically manifest machine at first. There are very many pre-formed reaction paths and sets of cooperating reactions and autocatalytic loops hidden in the network, all transparent at first. Then we let the few black dots on the left fill their outflux tubes, filling-up the next dots which in turn fill the next dot while being continually refilled themselves. While the water and CO_2 and some others on the left are kept at a constant level by the environment, their products are regenerated more and more slowly. The network of pre-formed reactions is slowly getting filled-in. This mental installation is lighting up like Olafur Eliasson's "Sun." This is ongoing creation and stupid eternal determinism in one. The scheme is very general. It is independent of the particular chemistry. It is a mathematical machine, which eventually includes Darwin's machine on the far right. I admit that I never worked it out as a piece of art, or in its totally detailed mathematics which would be the same thing. I was happy to see its existence. With the visible machine of life—in spite of its manifest beauty—turning out to be a virtually negligible feature within a mostly invisible machine that nonetheless can be seen with the mind.

This seems to be important to understand.

There is a set of initial substances which are just a few filled-in nodes on the left, but these substances are being steadily produced as "pool substances"—like water which is available in unlimited amounts on the earth, and many substances are continually swept-in by rivers which are emptying into the sea or maybe a lake at first or even Darwin's "warm little pond." There are some substances of constant supply. Then you have a situation where a few substances are in constant supply, out of which there can be produced other substances, and out of these again others and so forth. But the content of the slowly filling-in Teflon tubes does not necessarily become thinner and thinner. Sometimes there will be a feeding back—there are back-couplings. Pre-formed autocatalytic cycles and networks, if "initiated" by a first influx, can "reach back" to recruit an early (left-hand) source as a secondary self-replenishing source. The scheme describes a so-called "universal library" of chemical substances that can, in principle be produced from those on the left. They are almost all invisible at first that is, not yet produced. This invisible machine is the biggest analog computer that you can imagine, but it is not there in a hands-on physical sense, it is an invisible but nonetheless mathematically existing reality.

This is a graph to describe chemical reactions.

Yes, and the initial conditions are such that most variables are initially zero, only just a few like water and some carbon-containing molecules are already there, the famous most common ones in terrestrial life. "Chonsp" used to be an onomatopoetic word to help remember the most important chemical elements characterizing life. C, H, O, N, S and P have also been found on Mars, for example. These few are produced all the time, are being delivered free of charge all the time. But apart from the few pool substances in a non-zero steady state on the left, there is this huge network of potential reactions which happens to exist from the beginning also, albeit invisibly. The scheme gives you confidence that you can predict what comes out of this random network of pre-formed reactions. It turns out that you can prove that the network is evolving in a direction which generates more and more complicated, more and more efficient self-reproducing chemical sub-machines. Eventually, a "genetic machinery" is bound to get newly formed, that is filled-in. The evolution of the genetic code represents one such preformed thing that would happen in a gentle manner in a not too complicated way within the pre-formed network. Eventually, even what I call a "brain in the genome" is bound to be formed, that is, filled-in, but this is jumping ahead too fast.

The idea is a reduction of evolution to mathematics?

Precisely; you have a network of chemical reactions which is virtually infinite because there is an unlimited number of combinations involved a so-called universal library of chemical substances in the sense of Kurd Lasswitz. You can have C–C-backbones as we have them in life on earth—carbon is capable of producing chains. So you have chain formation included as a pre-formed product of the few initially present substances. Mathematically speaking the whole machine exists all the time while its visible elements follow what is called an "evolution." One can look at the laws of this evolution. It is especially fertile—hardworking—at places where autocatalysis and other over-regenerative cooperative loops are pre-formed. Such autocatalytic systems and larger cooperative conglomerates will be "recruited" along "shortest paths" in between them. Once self-reproducing larger—not too large—autonomous entities are "awoken" into non-zero self-maintaining concentrations, with modified (eventually "mutated") daughter systems being formed, that eventually involve a whole mutable "genetic" machinery, we have a system which I call an "Erdös growing automaton." I named it after Paul Erdös, but personally I only have "Erdös-number 4" (because a friend of mine with whom I have published a joint paper has Erdös-number 3). Such a growing automaton is very close to the mind of Erdös. The nice thing is that you have a free-of-charge giant computer here, much bigger than any artificial computer in the cosmos can or will ever become, but this computer is invisible even though it produces some visible substances of many sorts all of the time. Thus this represents a very big evolutionary principle, valid and at work within invisible machines that mathematically if not physically speaking exist with most of their elements. So evolution is not at all surprising for a mathematically bent mind. Evolution is pre-formed in the possible chemical reactions of but a few non-negative variables—substances that can combine to produce in a combinatorial fashion many, many other substances and networks of substances. The whole thing is, apart from Erdös, also closely related to Teilhard de Chardin with his idea of an "arrow" existing in nature. The basic theory behind this is thermodynamics—non-equilibrium thermodynamics. Ilya Prigogine had a similar idea. We met only later but I was aware of him since I was 22 or so, so he influenced my thinking. However, the idea just outlined is not explicit in his work as far as I am aware. It is a very general growing-automaton type of idea. Stuart Kauffman had a similar concept at about the same time. There is another scientist who also saw a connection between evolution and Erdös' growing automata, Joel Cohen, but this was several years later. I attended a talk of his in Vienna in 1977 and I told him afterwards that I had published virtually the same thoughts 3 years before. Unfortunately, he did not publish thereafter. So I learned the hard way that you should never tell someone else that you have been first because this could kill a whole promising cooperation. As a consequence, the whole field would not take off—the opposite of what a scientist desires. But of course, one can never predict the future.

Who organized the Vienna conference?

It was Ilya Prigone shortly after his Nobel Prize. What we saw is that dynamical systems theory is responsible for a predictive theory of evolution among non-negative chemical substances. It is important that their concentrations can never become negative. Otherwise the long "silent periods" would not exist and the whole thing could not become an "evolution." There are very long delays involved indeed, because only a few molecules are being built at first, and it then takes a long time until these molecules can, after being part of an autocatalytic network if lucky, become a source for the next type of molecules. It is an infinitely complicated network, but nevertheless one can look at the laws of such machines. Because of this scheme there is no doubt in my mind that "life" is unavoidable on chemical grounds in nature—not just in chemistry but also in nuclear chemistry.

Robert Forward wrote an astounding book, I read it a few decades later after being prompted by astrophysicist Hanns Ruder to read it, titled the "Dragon's Egg." It refers to a fictitious constellation of stars. In this piece of fiction, the egg is a supernova and leaves a neutron star in its wake. On this neutron star, nuclear-chemical life is evolving. The author then describes a fictitious communication in this scenario between human beings—astronomers using their instruments—and intelligent life on the neutron star. Forward is a scientist and the whole book is a bestselling science-fiction book, but at the same time it is absolutely correct in every scientific detail. It is a scientific breakthrough that he made by showing that the same type of evolution as we have it in chemistry exists also in nuclear chemistry since the latter enables equally many complicated reactions, only a million times faster—on the outer crust of a neutron star. So it indeed is a very general principle in nature that under open conditions—when there is free energy available as I forgot to mention at the beginning (but this was implicit)—you can predict life. Not just under liquid-water conditions as hold true on earth and maybe on Mars in a previous era, and no doubt presently on some of the ice-covered, internally liquid-water carrying moons of Jupiter or Saturn, Europa and Enceladus, respectively. But also in nuclear chemistry there predictably exists the same mathematically pre-formed but physically growing type of network producing a non-negative type of life. There might be other types of evolution which do not require non-negativity of the variables although nobody uttered such thoughts seriously. The whole thing is indeed an offspring of thermodynamics. Statistical thermodynamics—in the form of far-from-equilibrium thermodynamics—allows you to describe life, living organisms and evolution as an implication of the dissipation of free energy on the way towards the so-called "heat death" of the universe, as Rudolph Clausius taught in the second half of the nineteenth century. While energy becomes more and more dissipated, at the same time some fraction of the free energy is delegated to drawing-in more free energy into the system, to generate more complexity and improve the whole process. In this way, evolution is complexity-enhancing in an autocatalytic, self-improving fashion in a giant mathematically pre-formed system that if you wish is "living" even in its still colorless parts, as an invisible machine. One can hereby predict the invention of a genome that generates a "phenome." Actually, the generation of alternative "phenomes" hidden in the same genome is predictable, as C. H. Waddington, whom I met in Buffalo in 1969, first discovered in flies. One can then show that the genome becomes more and more intelligent. There predictably exists a "brain in the genome" riding on the back of the thermodynamic arrow. The arrow points towards more and more dissipation—entropy is increasing while the heat death is approaching—but on the way towards the heat death, you get this Teilhardian "complexification" (as he called it) in nature.

There is an old fairy tale that the devil can be employed to carry stones up the hill to build a cathedral. So it is on the back of the heat death that life is evolving. This is what thermodynamics allows, despite the fact that every free energy is dissipated eventually. On the way towards complete dissipation, there is complexification and this complexification is self-enhancing. This is life. So life would be a secondary effect within statistical thermodynamics. Very much later, it would occur to me that statistical thermodynamics stands not alone. There is a second fundamental discipline at work in which you have not a heat death as the driving motor, but rather the contrary, where complexification is the direct consequence of the underlying arrow. While the arrow of time in nature is so far believed to exist only on the back of an arrow that leads to dissipation and a heat death, in this new second theory, there resides a complexifying arrow that does not need to arise as a secondary effect but is the primary effect. There the main arrow leads to complexification directly. It is still virtually untouched territory, this sister discipline to thermodynamics.

You mean cryodynamics?

Cryodynamics, yes. The discovery of cryodynamics opens up a door that is much wider than the door of thermodynamics. Within the latter, there is this little niche in which we can have life. It could well prove important for the future development of science and humankind to more deeply understand cryodynamics because currently unfathomable implications might be waiting.

Could you say a word about the relationship between cryodynamics and Darwinian dynamics?

Darwinian dynamics is a part of dissipation while cryodynamics is intrinsically anti-dissipative.

Can you mix cryodynamics with evolutionary theory?

Both do co-exist, but their intermingling is very hard to understand. Early on, Lev Landau published a paper in the 1930s—the first author of the famous Landau-Lifschitz series of advanced physics textbooks—on the combination of repulsive and attractive particles in an open two-phase system. Later, Donald Lynden-Bell in England continued in that direction. He combined both repulsive and attractive forces in a calculational model describing a box with a membrane in between the two compartments, I believe. Both mentioned authors made their lives unnecessarily difficult in my eyes through combining attraction and repulsion and looking at the combination—a combination of thermodynamics and cryodynamics, as it were. This did not lead very far for starting out from too difficult a question. If you combine two disciplines of which one is not yet completely understood or even seen in its own right, forcing them into one combined little abstract model system, gives you no chance to progress very far. Therefore cryodynamics did not arise in its own right for more than a century after the high bloom of thermodynamics, despite the fact that these admirable scientists had glimpsed part of the essence already. The two sciences are primarily completely independent of one another. One should therefore look at them on different time scales, I feel. This could not possibly be seen at the beginning. The time scale of light flying through galaxies that are moving is a very long and slow one. A photon, passing by the Milky Way galaxy takes at least a hundred thousand years because the distances involved measure so many light years. Cryodynamics is mostly a very slow process in nature while evolution is very fast by comparison. On a neutron star, evolution is about a million times faster still with its nuclear chemical reactions than with our own chemistry (as Robert Forward stressed). So the two types of life so to speak—the two arrows, the one of thermodynamics and the one of cryodynamics—live on different time scales. If one realizes this, one can understand why the anti-entropic phenomena of the second arrow were not discovered before—it is because the time scale is so hugely different, and also because very few people are able to put such contradictory things into one picture in their mind. Roger Penrose, for one, did this in his book "The Emperor's New Mind" of 1989 when he drew two series of pictures, one placed above the other. One was the entropic arrow of dissipation, symbolically represented with particles getting spread-out more and more in a hand-drawn picture, the other underneath gave a similar rendering for gravitation, with an increase of "lumpiness" at each step, with but a few coalesced black holes remaining at the end. He clearly juxtaposed these two arrows. However, he could not possibly know yet—I believe—that there really is a new arrow in existence here that is just as powerful as that of thermodynamics itself. Nobody has a good intuition into cryodynamics so far.

A question on evolution theory; human beings are capable of deducing many things, how can evolutionary theory explain the existence of deduction in humans? Human beings can also love many divergent things. Can evolution theory explain the existence of this orientation in humans?

The simplified picture I gave you so far was—so one could say—essentially about well-stirred chemical systems. We all know well stirred chemical reactions like the Zhabotinsky reaction where you pour four liquids together. If you run it under well-stirred conditions, it can produce chemical waves in time—oscillations—, but we also can obtain chaos-so it can have all major kinds of dynamical behavior. Evolution in the way I described it was essentially evolution in a well-stirred soup whereby differences in space were not included. As soon as you include space explicitly, everything becomes much more complicated.

In reality, we only rarely have ordinary differential equations—the well-stirred case in fluids. We more often have "partial differential equations" which depend both on time and on space. There is the nice saying of Ulrich Wais, a friend of mine that "The Lord has created many animals and plants and many partial differential equations." Partial differential equations are much more difficult to handle than ordinary differential equations. This model of evolution that I gave you was based on ordinary differential equations. So if we add space to the whole thing, everything becomes more complicated in a sense. However, in other ways it becomes simpler because one can now have a cell which has a membrane, and we are familiar with all this in biological evolution. It is just somewhat more difficult to handle didactically—in a simulation, for example. In a few decades from now, if the planet is still there, computers will be so fast that one can do simulations of all the things we were discussing, but to come back to your question. There is more to say about space in the context of pre-biochemical and biochemical evolution. Survival can no longer be defined only in a temporal fashion. Darwinian evolution, natural selection, does not include space explicitly. It concerns chemical reactions, which are based on partial or ordinary differential equations. The stochastic versions of the latter needed at low concentrations are even more difficult.

As soon as you have an animal as the fruit of evolution, this animal lives in an environment in which its survival depends not just on chemical conditions that are more or less well stirred locally, but also on where it is at a given moment in time, how the conditions for survival are at the special location of the moment. Plants can only thrive if there is enough light, and underneath the big trees in a forest there is not much light left, for example. So space plays a big role in biological evolution. But there is a different aspect of space which comes into sight here: fast changes of survival as a function of position in space. Plants only make use of space with their roots and phyllotaxis and so on, but they usually cannot move. As soon as an organism invents movement, locomotion, suddenly space plays a major role for survival. It is a second type of theoretical biology that emerges here if you allow that survival depends not just on history and chemistry and partial differential equation-type situations, but also on the momentary position in space of the whole organism. This is the problem of locomotion, or rather the problem of "behavior" as it is called.

Behavior is something new in biology which has nothing to do in principle with the biochemical underpinnings of the same organism. You can pose the question of "survival conditions" not purely in time, as Darwin did when he predicted his mechanism of natural selection (which eventually allows you to even predict the "brain in the genome"—we talked about this before). Survival depends not just on history and chemistry, but also and very acutely on position. "Position-in-space dependent survival" is a totally different problem. It is still a mathematical problem as in chemical evolution, but it is a much simpler problem because it proves to be a well-defined mathematical problem that is independent of the myriad historical chemical accidents and their relics of biochemical evolution.

In evolution theory we suppose to pass through many steps, through many animals to arrive at human beings. Humans are able to work with mathematics and philosophy and they can produce many systems of logic. Can evolution theory explain the jump from this animal point you described up to a theory of this more modern type of brain where at the end we have a human being?

What is the "place of man in the cosmos"—Max Scheler asked in 1926.

I mean the possession of the system of logic and also the possession of love. In animals, love seems to be of not so high a level, in humans, many different levels of love are observed. Can evolution theory explain this?

This is the most difficult problem in our context, the difference between humans and animals. Presently we are still at the point where we try to predict the brain and thereby understand it. Both animals and human beings have brains, but human beings make a different use of their brains than animals do, as far as the evidence goes, and this needs to be explained as you say. And it indeed can be explained. It is a bit easier for me if I first explain how the brain works. Let me return to this point first if I may.

The brain is needed because survival in space no longer depends on prehistory and well-stirred chemical conditions alone, but acutely on position in space. This problem proves to be decoupled from the chemical history and prehistory. It turns out to be a well-posed "optimality problem" in the sense of mathematics. It is called the "travelling salesman with alarm clocks problem." Now I should first explain what the travelling salesman problem is. Suddenly we are here right in the midst of pure mathematics again. It is a very nice, also economically important, problem and one that has also attracted the interest of physicists. For example, my famous physicist friend John Wheeler wrote a long paper on the travelling salesman problem. It is a very simple question to be solved. You have a number of towns, say five, and you have a salesman who has a car available and has to visit these towns to sell something. The simplest assumption is that he has to visit every town only once and that he wants to save fuel while visiting all the towns. If you have five towns, it depends on how they are located to determine what the shortest path to visit all of them is. This is the "travelling salesman problem." This simple problem is mathematically very difficult and interesting, because if you increase the number of towns, then to find the shortest path of connecting them all along a single route becomes more and more difficult in a very rapidly rising fashion. The complexity goes up, as one says, in a "non-polynomial" fashion which is another way of saying "exponentially." The complexity—the number of trials necessary to find the optimal solution—grows very, very fast, more than exponentially fast. So the complexity of the problem came to be called "non-polynomial complete" or NP complete. Nonpolynomial is "exponential" and "complete" means that very many other problems have just the same structure, so solving one means that all of them can be solved. NP-completeness is currently a famous conjecture in mathematics. If your algorithm runs on a quantum computer in polynomial time, you will become the richest person on earth. Most probably, I would guess, even a quantum computer cannot solve an NP-complete problem but this is still open.

Why do I mention this strange phenomenon of the travelling salesman problem in an abstract landscape? It is not just mathematically nice. There exists a cousin to this problem which is called the "travelling salesman with alarm clocks problem." It is slightly different—you do not have to minimize mileage as before, you only have to avoid that an alarm-clock is going off. Here, the travelling salesman is carrying a bag with a few alarm clocks, and each goes off after a certain fixed time unless he finds a town or filling station of matching type. All you need is imagine a car which has not just one tank to fill up with gasoline or electricity, but needs to visit other filling stations as well—say for oil or for new tires or for repair and these have their own specific filling stations scattered in space. If you do not make it, the car is removed from the game and the journey is over. Each of the fuel tanks of different colors is equally vital. Hereby they can be assumed to have equal time limits without change of generality. Mathematically you again have a travelling-salesman problem, only with the "constraint" being changed. You may have five alarm clocks, for example, or just three. Even then the problem is maximally difficult. You have to make sure that you visit filling stations of different colors in enough time to avoid the alarm clocks being set off. The filling stations are—for example—randomly distributed in space. No colored tank must ever go empty, otherwise you are dead. It is a very simplified but realistic version of survival in space when you have sources that need to be visited from time to time like a source to drink, a source to eat and a source of shelter in the night. One for every different "biological duty", as it were.

When I was a child and got restless, it could happen that my father would ask me whether I had not to "fulfill a biological duty"—to go to the john. We know we have to drink and to eat, and so on. This travelling salesman problem of the alarm-clocks type is a very general mathematical problem which captures the essence of a type of predicament that we are all familiar with. It is not just humans that are familiar with this predicament to survive by finding certain goods. Otherwise, higher biological organisms would no longer exist. One of the types of sources would consist in finding a partner for procreation. This second type of travelling salesman problem is mathematically speaking a well-posed problem again. The solutions can be studied in abstracto, being "NP-complete" again. You will need a computer bigger than the visible universe to solve it if the survival time is to be maximally long under a condition where the time gap between the maximum travelling radius per tank filling and the minimal temporal traveling distance between filling stations of the same type becomes small. When you have very much time and hence many occasions to fill up a particular on-board gas tank, you do not need to care much because you so frequently pass by one. Garey and Johnson independently discovered this travelling salesman problem. And Eric Charnov also saw it implicitly in ecology under the name "optimum foraging theory" in the same year in which I first proposed it. This is a unifying problem which must be solved in biology—in any biology on any planet including the "Dragon's Egg" neutron star of Robert Forward.

Now to the point: This well posed mathematical problem needs to be solved by on-board machines called "brains." To solve this problem is the very task of brains and the reason why brains are needed in biology. In other words, you can predict the necessity of brains because the travelling-salesman-with-alarm-clocks problem exists in any biology in which survival depends on position in space. Darwin was probably quite aware of this—a question for the historians.

This is not metabolic adaptation?

No, that is the point. We now have two types of adaptation; "metabolic adaptation" toward the physico-chemical environment on the one hand—which is taken care of by ordinary chemical and biological evolution—and "positional adaptation" on the other. Metabolic adaptation is Darwin and Wallace's adaptation. "Positional adaptation" represents a second type of fundamental biological problem which needs to be solved too, and which unlike metabolic adaptation happens to be a well-posed optimality problem in mathematics. You need nothing but mathematics to solve it, and for this reason you can predict the necessary properties of "brains." Hence brains are much easier to understand than the rest of the body in biology because you do not need biology at all for describing their functional features. In all biologies in the universe, it is always the same positional-adaptation problem that has to be solved and is being solved. Because of this mathematical well-posedness of the problem, you can try to find a "brain equation." The latter is an equation which solves this travelling-salesman-with-alarmclocks problem up to a certain level of efficiency. Solving it completely will demand a computer as big as the universe. So, this NP-complete problem is the basis of a "theoretical biology of brains" in the same sense as there is a "theoretical physics of gases," for example. This "hard" theoretical biology of brains parallels theoretical physics in its self-understanding. It can become the basis of a love theory.

My biochemistry teacher Günter Weitzel used to claim that "all life is chemistry." We can now say in the same vein "every brain is mathematics." It solves a mathematical problem which is well-defined and can be dealt with the usual instruments of math. The modern science of "bio-informatics" contains the brain equation at its core. For some reason, the specialists presently treat it only as a special exercise in undergraduate courses, if at all. I apologetically would associate it with "half" of the task of theoretical biology as Ludwig von Bertalanffy defined it. The other half would be the not yet elaborated theory of the "brain in the genome" defined 40 years ago with Michael Conrad. This is "music of the future," of course. No one believes this so far, I am afraid. Konrad Lorenz helped me see this. I remember when we talked shop as we did about ten times in 1966, how the question came up. Or rather, the underlying phenomenon came up—of crabs living at the seashore in the region of the tides. There the water is going up and down every few hours. So these animals should take into account in what phase of their environment, of its physical rhythm, they are living at every moment. We realized that this is not something that can be genetically programmed-in in Darwinism. Genetic adaptation cannot predict what at a certain time on a certain day at a particular seashore the position of the tide will be. It would be impossible to be adapted to such an environment on the basis of genetic information alone. This was the first instance which allowed us to realize that genetic information and evolution theory in the Darwinian sense cannot cover this problem, namely the problem of "positional adaptation" in space and time, as I call it. It represents a question of its own in theoretical biology.

You mean that the brain always computes the exact position?

This is what I am driving at. The simplest case is that your organism formed by chemical and Darwinian evolution must be cued to doing something adaptive, like switching-on a different breathing mechanism in time before the high phase of the tide comes. For anything like this, one can predict that you need what is called a sensor. Only if they have a sensor can these biological organisms survive—unless their physiology is made universally resistant to all these changes by metabolic genetic adaptation. Without a sensor to obtain information that is not part of the genetic information, this system cannot behave in a temporally adapted fashion. So there predictably exists a second type of survival-relevant information that goes beyond genetic information. That was the insight obtained jointly in 1966. We soon saw, of course, that it is not just temporal information that is vital in this sense, it is also spatial information. If you are in the wrong place, you will be eaten, for example. So it is both position-in-time and position-in-space. The information about space is even more important in general. It is "action-specific readinesses"—moods—that are needed in this context. It took me seven more years to arrive at the brain equation. Mathematically, the situation that one arrives at is the travelling salesman with alarm clocks problem. Hereby a biological system needs to keep track of how much time has passed since one of its different "tanks" has been filled and hence now needs to be filled-up. As a theoretician you realize that you are free to assume that all the tanks can have the same time limits to be refilled—it makes no difference for the difficulty of the task. In reality, it is different—we cannot stop breathing except for one or a few minutes, for example, but we can stop eating for a few days or weeks. These life-deciding clocks are not equal in reality. Only mathematically this is no major determinant of the complexity of the problem to be solved. Then you can start trying to solve this problem.

There is a book by Konrad Lorenz of 1973 titled (in literal translation) "The Flipside of the Mirror." In English it has the title "Behind the Mirror" which is misleading. What he meant with the word flipside (in German Rückseite) was that we are endowed with this "mirror" in which the world appears to us, the mirror of consciousness. That the world is something visible like in a mirror, but that behind this mirror in which the world appears to us, we have a machine as the flipside of the latter. In case something goes wrong, the driver needs the help of a mechanic who is ready to make his face dirty going underneath the machine. We are doing this right here, we are not afraid to use the screwdrivers of math to do justice to the machine involved. The deductive science of behavior tries to understand the workings of this machine which for us is so all-important. It is not a chemical-reactions type machine, as in the case of our body and our living cells, it is a brain-type machine. Even if its cells' chemical constituents were to a large part suddenly destroyed by ionizing radiation to be "metabolically dead," the machine's electrical and mathematical behavior along with the attached consciousness could still go on for a while—in a dead organism. So the brain is not alive in the biological sense if you wish, it is a different type of machine with a different type of life, even though it uses the same cellular elements, the same wet-ware. You can replace the living elements of the brain by computer elements, as philosopher Hilary Putnam seriously proposed in his 1981 book "Reason, Truth and History." We know from experience that computers can be intelligent, but it was not known before that we are computers as far as our brain life and consciousness is concerned. This type of behavior and what we are doing with our brains, all animals do, is something which belongs into informatics, not into biology. Only if you want to repair it, it is biochemical elements that need to be repaired in general so far. "Brainology" is a completely different science of its own even though biologists are legitimately interested in both types of life, biochemical life and mathematical life.

Chapter 11
Thermodynamics and Its Proposed Counterpart

Let us try to explain how the universe evolved. After our talking about biological evolution, let me ask, can we speak of a Darwinian dynamics in cosmology?

In science we try to have a complete picture of everything when we look at it from an external point of view. A complete picture is the goal. The strange thing is that we then get stuck with a particular "cut" that we are living in at a given moment in time, the fact that our consciousness is attached to a specific body, and so on. So there are two different views of the same reality. It is much easier to take the detached or exo perspective to talk objectively about the whole thing. This is the main aim of physical theories—to try and understand the whole thing which includes consciousness.

First, I would like to stick to the idea of a machine seen from the outside; the cosmos is a big machine that we try to understand by deciphering its formal laws, as Newton first did successfully in the footsteps of Kepler and Descartes, and as Einstein continued. In a previous discussion we saw that there is not just the arrow of entropy increase in thermodynamics, as is well known, but that there is a "second arrow" at work in the universe, the "ectropic" arrow of cryodynamics. If the theory is correct, the whole cosmos is an eternally recycling machine in the sense first seen by Heraclitus and then later on by other people, most lately Poincaré and Boltzmann. Nietzsche—who planned a new career studying physics in Paris—was the philosopher of eternal recurrence in the 19th century. An infinite recurrence time means that the whole thing would have to be deterministic in the sense of repeating the same thing all over again. The assumption of infinity is so big that it all will remain full of surprises for an eternity and never become boring. In quantum mechanics, there are modern theories which claim that whenever the wave function is reduced to give a particular result at one point in the universe, the wave function of the universe is split-up in almost infinitely many ways so that all solutions of Everett are real in some branch. This assumption, by the way, implies a sort of immortality. It is very confusing to Europeans that here one suddenly finds oneself in a typically Eastern, Indian way of thinking. Everett himself was—as John Bell found out—actually quite close in his thinking to that of another

A. Sanayei and O. E. Rössler, *Chaotic Harmony*,
Emergence, Complexity and Computation 11, DOI: 10.1007/978-3-319-06781-0_11,

friend of John Bell's, the Dalai Lama. The decisions made in quantum mechanics—as to which world is going to be actualized at a given moment in time by a given measurement—is a decision that varies from moment to moment. This means that one is living in a different world at every moment without noticing the change. Bell wrote a paper about this, titled "Quantum mechanics for cosmologists," which indirectly refers to Everett's theory because Everett's is the only version of quantum mechanics compatible with cosmology; since the other versions by definition require a measuring apparatus larger in size than the object to be measured, which is impossible if the object is the cosmos. In his paper, Bell proposed that the different cuts—the Everett-cuts of the universe—are not, as Everett had proposed, simultaneous so that the many observer-specific worlds exist in parallel—but rather mutually successive. So, one would live at every moment in a different Everett world. He drew the connection to his friend, the Dalai Lama. He is believed to continue after death in a different life and body until he eventually decides to escape into Nirvana which is seen as the ultimate improvement. It is controversial whether the Nirvana of the Indian enlightenment is a form of permanent death or permanent bliss; at least I am not sure which of the two interpretations is the more sensible one.

To return to physics, we have the idea of a giant "world machine" which stems formally from Descartes and Newton. This world machine should or can be assumed to last for an eternal duration. In physics we have two versions of thermodynamics, as is not very well known. On the one hand, there is thermodynamics itself, on the other, there is cryodynamics. Due to their interdigitation, the whole system called cosmos can become eternally recurrent. This is a very satisfying global picture which only became possible after the discovery of cryodynamics. It is very important to falsify cryodynamics because if validated, it opens up the window towards a totally different cosmology—a prospect that is completely against the grain of the last 84 years of cosmology which in this light looks very dogmatic. Evolution would be part of the new theory as a consequence of the dissipative half of physics called thermodynamics, while cryodynamics would be "antidissipative" in its nature. This "doubling" of thermodynamics as it were would enable an infinite cosmology in space as well as in time, in contrast to the short period of a few billion years allotted to us by the current "pre-cryodynamic" cosmology, if I may call it so.

Namely 13.8 billion years?

Right, 13.82 billion years at the current moment in time, but this number changes almost every year. One should never be dogmatically critical of what everybody is thinking, but one should nonetheless always try to find a more open window, so to speak. The new cryodynamical window suddenly allows one to return to the idea of an eternal recycling, around-motion, perichoresis. I talked about this recently with philosopher Friedrich Kümmel in Tübingen. Heraklitus said at about 500 B.C.: "metabállon anapaúetai"—metabolizing it takes a rest, that is, it remains unchanged. But if one is really so optimistic as to think that a complete picture of physics is possible as I would support, there is this big nuisance of quantum mechanics on the one side, and on the other side there is the other big nuisance of consciousness. Why are we attached to one particular time slice cut through the universe, and to one particular body populating it? Hence the "Exo" physics we just talked about has to be complemented by "Endo" physics—the physics of assignment.

Why am I assigned this consciousness or any consciousness at all? We are in the possession of consciousness as the only substance that we can be sure about, and that constitutes us. This consciousness has no place in physics. One can prove that it does not exist at all (were there not the direct evidence). Then we have to try and make a connection between consciousness as we experience it at this given moment in time and the question why it is given to a particular inhabitant of a given body, at a given time, in the vast Now-less world machine. We have to find the point of contact between these two realities, the reality of the world machine on the one hand and the reality of the momentary consciousness on the other. This assignment is an "add-on" characterized and guaranteed by the fact that we also are provided with color—like the colored fruits here in front of my eyes. So both COLOR (one should always write it with capital letters) and the NOW in assignment are the real riddles within this big originally Newtonian picture of physics. One might think this is psychology entering surreptitiously here, but this has nothing to do with psychology as a discipline. It is the "miracle"—the fact that something is injected into the physical machine for which we cannot find any reason within physics and hence of course also no explanation. Why my consciousness at this moment? There is nothing in the laws of physics prescribing this. Thus we are subjected to a purely metaphysical intrusion which nonetheless is absolutely compatible with physical thinking. In physics, this is called assignment—an "assignment condition."

Newton first introduced two conditions in describing the world as his basic insights: the "Laws" of the world machine on the one hand and the "Initial Conditions" (I.C.) on the other. These define which particular state the machine is in at a particular moment in time. In addition to these Big Two, we have to introduce a third equally basic element, the "Assignment Conditions" (A.C.). They describe where, within this big machine of the universe, a particular consciousness is attached, or more properly speaking finds itself attached to a particular element of the big machine. For if we are honest with ourselves, all that we have is this consciousness, and if we are lucky, we can find within the waking dream of consciousness a picture of the world as a consistent machine. We can then ask the question of the docking point in the world machine and within it in the body machine where the assignment is implemented. In older times, one would have called this type of thinking "religious thinking" because if one believes that there is a reason for the assignment at this very moment, one is suddenly confronted with accepting-or-not a power that lies beyond physics and is responsible for this dream of consciousness as it is being played to my soul at this moment.

A cell-phone type analogy if you wish. Please, interrupt me so I can escape again from this swerve towards the frightening subjective side of physics.

Let us return to Darwinian dynamics and cosmology. Is there a type of natural selection in cosmology?

Now we are back in physics. If we try to understand biology, we need this element of natural selection discovered by Darwin which can be explained in terms of chemical evolution, as already seen by him. We have within thermodynamics the phenomenon of entropy generation. Within the latter we have the fact that part of the increase of entropy can be used to build up more complicated machines like flames and organisms or "dissipative structures" as my late friend Ilya Prigogine called them. This is a question purely within the scope of physics. We can explain it in the "thermodynamics of irreversible processes." Recently, the second 50 % of physical dynamics governed by "cryodynamics" was added.

You think in cryodynamics, the procedure would be inverted? Does this mean that the arrow of time is changed there?

It is another arrow of time that cryodynamics brings in, it coexists with that of thermodynamics. Bringing the two theories together is currently a far-from-solved problem. There are only some first ideas about how to proceed. Cryodynamics guarantees an infinitely long existing cosmos which then produces Darwinism, that is to say, allows for dissipative structures on the thermodynamic side which culminate in biological evolution. This happens on many celestial bodies on the basis of different chain-building chemistries and also on the basis of nuclear chemistries on neutron stars on a much faster time scale, as Robert Forward first saw. Biology is an element of this arrow, Teilhard's arrow. Cryodynamics in turn regenerates the conditions that evolution can happen again and again because it makes sure that entropy is not the only king of the universe. There is a second king or queen; that of "ectropy generation" in cryodynamics. I realize that I do not yet have a simple composite picture gluing these two theories together. Cryodynamics is not a competitor to the irreversible dynamics of evolution theory. It rather is a precondition in order for evolution to be able to start infinitely often in the cosmos. Evolution contains the attractor of Teilhard de Chardin, the "point Omega" towards which all life forms in the cosmos are converging in an asymptotic fashion. On the other hand, in a sense cryodynamics is much more powerful. It does not need to produce far-from-equilibrium structures at the expense of much more dissipation going on. While in thermodynamics organized structures incorporate "negative entropy" (as Erwin Schrödinger called it in his book "What Is Life?"), organized structures arise directly in cryodynamics. If you accept the picture, in thermodynamics life is something like a crowning cream on the coffee, a side product of the thermodynamic arrow of time on the way towards the "heat death" (Wärmetod), as Clausius, inventor of the word "entropy," envisioned the future of the cosmos. Life is something like the last savings before eventually the heat death destroys all self-organization. However, we now have an antidote, something like a cosmic anti-death, a bit like "The spy who came from the cold."

There is an evolution, too, on this "left-hand side of physics" as it were, in cryodynamics. It is not yet as transparent to science as evolution in thermodynamics is. In cryodynamics, we get an ectropic arrow that is the main player from the beginning, unlike the arrow of evolution based on the opposite process of entropic energy-dissipation. That is, entropy is not increasing along with a bit of order created as a side effect, but rather the order—ectropy—is directly generated as the main effect. It is only very hard to see it in action in its purest and most powerful form—and capture this in a grandiose picture.

Biological evolution can be considered as a chain. Every link in this chain is transparent to us. My question is; if you want to apply the Darwinian viewpoint of evolution theory to cosmology, what is the chain there and what are the chain links, so that we again have a causality and every chain link is responsible for the next one?

The Darwinian and Teilhardian viewpoint is in principle understood. Human beings like to have a theory that is causally understandable everywhere. Quantum mechanics could be fitted in, with the notion of endophysics making this possible in a deterministic universe. So the machine theory of Descartes still stands behind the completed picture that is in principle obtainable. In the full picture, we suddenly find a second part, cryodynamics, added to the familiar thermodynamic part. However, the whole thing is not fully understood yet—if it is indeed true that there is an ectropic arrow at work in cryodynamics. We would then have an arrow that is a competitor to evolution, but a competitor that is much stronger because evolution is as we saw only a sub-element of an arrow that tends towards a universal heat death.

We have life like a little wavelet crowning an ocean, like an i-dot added to a capital letter "I" that in principle could be neglected. But with cryodynamics, we now have the capital letter "I" itself to ponder. The letter itself is ectropic in cryodynamics, not just an appendix of it is ectropic. There exists a whole new creativity in cryodynamics. It is much more direct than one might think. Capturing the prototype process is still elusive. Nevertheless, finding something like individual intelligences within cryodynamics is to me impossible to picture at the moment. This is "terra incognita." The scientific community for once needs help from the competition—from people who are dealing with the mind—to get an idea whether something like consciousness could have emerged more directly, in the new ballpark of cryodynamics. Another obstacle is that cryodynamics "lives" on a much slower time scale than the fast time scale of evolution which is just a few billion years old, a number worth nothing compared to infinity. So there might be something like angels standing at the end of evolution, but there also might be consciousness-endowed beings within the cryodynamical branch of the universe. Thermodynamic evolution exists only as long as we have free energy flowing-in. The cryodynamic part suspends this limit, it eventually produces very many black holes which in one model re-circulate 50 % of their population in an eternal metabolism. Since black holes are never finished, this recirculation is possible in principle. In this case the thermodynamic part of the dynamics of the universe can also go on forever. So much for the combined picture, so far. Whether or not something analogous to life and/or forms of intelligence can arise within the new competitive arrow to thermodynamics itself since this theory is just as big and complex as thermodynamics, is open. This is a very exciting and to me at the moment almost frightening question. For it shows that half of physics, half of natural science, remains yet to be understood. If we today have wonderful machines based on thermodynamics, we will have even more wonderful machines based on cryodynamics eventually, because the newly discovered 50 percent of the full dynamics promise more than just a parallel "second business." It is rare in science that something as large as one of the biggest traditional fields is added to the ballpark. Cryodynamics appears to be a new country.

... new continent. If we consider cryodynamics and look at evolution: Would a "reverse evolution" from human to monkey now be possible?

Within the traditional continent, we have biology, into which human beings belong. On this continent—if we now stick to it—one can understand humans as the products of a very rare accident that happened in evolution and enables a function change to occur on the ontogenetic level of the individual. So we have now slipped into another shoe as it were. For humans make a radically different use of their own brain compared to the use that an ordinary well-behaved animal makes. Animal brains are controlled by evolution, they are not persons. This is another new dichotomy in science—that one has to distinguish between intelligences that are persons and intelligences that are not persons. As far as we know, there is but a single person-type intelligence in the universe, the human person. There is this theory which I developed in the aftermath of conversations with Konrad Lorenz: That one can have a radical change in the use made of a highly developed brain such that the owner of this brain becomes a person. This is a bifurcation of its own, a function change on the epigenetic level during the lifetime of the individual that ordinarily occurs at a very early age. Nature as controlled by evolution is "autistic" everywhere. I call this physiological autism. The only example of non-autism is provided by human beings so far. One can explain why human beings have become non-autistic. If I may repeat from an earlier session; it has to do with the experience of benevolence made by the toddler as it is being shown to him by the mother. You can have a system which has a certain type of functioning—runs on an attractor—and if you then change the initial conditions of the system, the same system with the same hardware under the same parametric settings can be switched into a different basin of attraction, a different mode of functioning. This is the theory of "function change" as Bob Rosen called it; another word for this general phenomenon is "hysteresis." The person attractor is a new type of attractor that is compatible with the functioning of a biological brain of sufficient complexity. But it does not manifest itself ordinarily in other species. Human beings are mirror competent bonding animals with a special social coupling. Other bonding animals fulfill only one or two preconditions for this function change. My publications on this since 1968 persuaded no therapist to try it out in childhood autism so far. It is so frighteningly powerful if correct. You then can heal from autism not only young human beings, but also a young orangutan, for example. He or she would then become a person, too. She or he would even be a more highly brained person, as I saw lately. Today, in the age of "artificial intelligence," everyone is looking forward to a very highly brained artificial intelligence; I love Steven Spielberg's movie "A.I." It should be possible today to get the same miracle already from biological hardware—like that of an orca whale or an orangutan—is so frightening a prospect that even the professional therapists for autism in human beings are unable to believe in their own power of healing. Of course, human society might have a big problem integrating not just different genders, as to date, but also Chewbaccas—George Lucas' creation in his "Star Wars" movies. This orangutan was the orange-coated superhuman pilot of the flying saucers. Science fiction authors have solved this problem long ago and the public knows about it. As seen by Leo Szilard there could be a dolphin who can talk and be wiser than any human because his brain is able to understand bigger objectively existing connections than we poor medium-brainers can—but this a different topic.

Evolution theorists try to explain how the universe or parts of the universe evolve. On the other hand there is a theory in physics, quantum mechanics, which has the same mission—to explain everything in the evolution of the universe. Now my question; is there any bridge between these two sciences? Do they contradict each other or can we tie them together?

This is a very difficult question. Physicists doing quantum mechanics are convinced they have the more basic theory. Accordingly, the older classical theory is regarded as an epiphenomenon of the quantum reality. There is only one competing way of looking at physical reality, called endophysics. My friend David Finkelstein coined the word "endophysics" which means that you can have a classical theory of the universe which implies quantum mechanics on the inside as it were—endo means on the inside. Endophysics is now 30 years old. Endophysics assumes that the world is classical at the basis and that as an implication of this classically-causal machinery you have an endo perspective which contains quantum mechanics. I submitted it in a paper sent to Science magazine as mentioned. The first sentence of the paper read "A new science is described." Although I was aware that they most likely would reject it—it was later published by John Casti in a conference volume—I felt obliged to submit it out of fairness. It is an irony that if you really create a new science, as I was blessed with in cooperation with David Finkelstein—, the scientific establishment is predictably opposed to it. If the comparison were allowed I would say this experience was made in the opposite direction by Max Planck a century ago when he gave up on classical physics. Even the automobile was not accepted at first—until a woman, Berta Benz, drove the first 100 km in the 4-wheel motor bicycle that her husband had built but hesitated to take seriously himself. This must be even more true if new types of technology promise to stand at the end. I admit that I very immodestly called the foreseeable products "world change machines." The latter would change, not just something within the world, but the whole world. This is even more frightening than cryodynamics which is a tame pussycat by comparison. Cryodynamics belongs into the usual exo picture of physics, adding just some 50 % to it. But for 30 unsellable years endophysics is still more subversive in the sense of freeing us from the trodden path.

Do you see endophysics as a bridge between quantum mechanics and evolutionary physical theory including cryodynamics?

We indeed can basically have a classical theory—one that is both cryo and thermo, so to speak—and then within this classical theory, we can look for the interface formed for internal observers, with quantum mechanics forming an implication.

This is not a very new question. In quantum mechanics, we have the Copenhagen interpretation and on the other hand, we also have the Everett interpretation. Neither explains which quantum states become in effect classical states.

You mean, why in a measurement process a certain result comes out.

We have for example physicists like Wojciech Zurek who wants to build a bridge between quantum mechanics and evolution theory. He calls it "quantum Darwinism." Using this mixture, he attempts to explain how one quantum state becomes practically classical on the level of our macro experience. He takes the environment into consideration and introduces his key notion of a "non-quantum superposition state." They are not created equal, some being more apt to survive in one particular environment than the others. Actually, I see some formal problems here because I cannot understand the meaning of "environment" as it is used. He considers the term "environment" in the way we know it in daily life, but "environment" can also refer to a quantum state that has become an element of the latter. So there is a gap here when trying to make a bridge between quantum mechanics and classical evolution theory. What do you think of this problem? Can quantum mechanics be improved so as to explain evolution theory as well?

All of this is very interesting, but it presupposes that the exo/endo distinction which I introduced does not exist. It takes for granted that quantum mechanics is the basic theory of the universe. From there, Wojciech tries to understand evolution—you here have a lower level of evolution than in the case of Darwin, the level on which the quantum states are selected. I have much sympathy with this kind of approach, but my alternative way of cutting the world into pieces—namely into two layers, the exo layer and the endo layer—allows me to avoid this problem. Quantum mechanics would be just the endo aspect of a classical theory—and so the classical theory would be the most important theory. Quantum mechanics could be derived from this classical theory. I apologize that endophysics asks for a different way of interpreting the same phenomena that are so important to understand.

Is the main idea of quantum Darwinism—that you never do any direct measurement on anything and that the environment acts as a witness or as a communication channel or co-determinator—to your liking?

In a sense, I admire this approach. But I also see it as a misunderstanding of Everett's theory. I may be wrong here because Everett never talked about the environment.

I don't understand; what is the meaning of environment in this context? I think the notion needs a very precise definition. Here we have a quantum state which has become classical. Can we understand in this way how a quantum state comes to a classical representation?

I do not think so. If you want to continue here, you have to explain everything in terms of quantum mechanics. Then also the environment would become a well-defined quantum reality.

You mean an environment for everything?

This is what one has to accomplish if possible. I mean that this is what is being attempted by Wojceich. But Everett would probably say that he has already taken care of this—it is a pity that we cannot ask Everett himself.

I think we can finish this chapter on evolution theory. We talked about different areas and also tried to build a bridge between quantum mechanics and evolution theory. However, I think that many of the questions I asked are open problems for research.

It is wonderful that so many open problems exist.

I think we sorely need paradoxes and also some open problems for progress in science.

Chapter 12
Newton's Laws, Symmetry, Noether Theorem

In the previous session, we talked about evolution. Another subject with the potential for one or two sessions is symmetry. Let us talk about symmetry in nature, symmetry in mathematics and symmetry in physical laws. At first, let me ask you: How do you define symmetry? And secondly, what are your thoughts on symmetry in physical laws? Can you, please, give a definition of symmetry in general?

I have to admit that I never thought of defining symmetry in general so far. I know there are some mathematical ways to do it. It is a fairly intuitive concept, so we can rely on looking at a special case like mirror symmetry as an example. What we see in a flat mirror is exactly the same thing as what we see in front of it apart from its being flipped in chirality. It is of course very nice and challenging to put this insight of "equality" into mathematical terms. If you have equations for example, it is very easy to see in general that if you make a change of sign, it is the same thing with a change of symmetry, but there are more subtle symmetries. I was very much, I would almost say, moved by a book by Newton. It is called "The Leibniz-Clarke Correspondence" and was published in 1717. Half of its content is written by Leibniz and the other half by Clarke who was a friend and confidant of Newton's. So this part is effectively Newton's. It is a correspondence held, one year before Leibniz passed away, between these two towering mathematicians. It describes in some chapters a type of symmetry which is not very well known. Leibniz argues that it is conceivable to have the same universe twice; what would happen then? It is clearly allowed to consider this symmetry. He claimed that if the whole universe existed twice, this would make no difference compared to its existing only once. Newton was taken aback by this idea, Leibniz had to explain. He claimed that if you have exactly the same total universe twice, you still have only one universe because they are identical in every respect. He went on to claim that even if you invert the orientation of one of the two otherwise identical specimens, you still have only one universe since the copy is identical in every respect—when you are inside of it, for example. So if you have two mirror symmetric universes, it still is only one universe—a very strange idea of Leibniz' indeed. Then he went on further to include

A. Sanayei and O. E. Rössler, *Chaotic Harmony*,
Emergence, Complexity and Computation 11, DOI: 10.1007/978-3-319-06781-0_12,

the assumption that the two mirror-symmetric universes border on each other in such a way that there is a plane in between them belonging to both half universes, so that the plane would look like a mirror. Then there would still only exist one half of the combination. What would then happen if you threw a glove or some other object that you assume could pass right through an identical copy—like two soliton waves—into that seeming mirror? You guess it. The glove would go to this surface to disappear through it while simultaneously a mirror copy of the glove would come out through the same surface. Note that there is only one universe. The glove would return from this transparent wall, reflected but its chirality—its handedness—would have been changed along. So would that of a soft enough molecule thrown against that "mirror." This was quite unusual an idea of Leibniz at the time. It was very radical because he also claimed that symmetry is a feature in mathematics and reality that is not just an infinitely accurate property, but a transfinitely accurate one. Like the modern idea of fractals—where we have this Mandelbrot type thinking of everything repeating itself all over again not just infinitely many times, but in such a way that the surface around it is transfinitely large. This is similar to the circumference of a von-Koch snowflake. You know the von-Koch snowflake? It starts out with an equilateral triangle, then the middle third of every side is expanded into an equilateral triangle again (without bottom line), and so forth, so that eventually the length of the von Koch snowflake is infinite—and even transfinite. For we have a finite number, 4 over 3, by which the overall circumference is multiplied at every step of putting a little hat in the middle of every side. Four thirds to the infinite, the new length, is the same as two to the infinite, times the original length. Two to the infinite is a transfinite number, it is called Aleph-one. Symmetry like fractality is a transfinitely exact notion in mathematics. Symmetry is an assumption of transfinite exactitude made about an object. In this sense, symmetry is a very, very strong clue to understanding nature and physics. If one finds a symmetry that deserves the name, it is a constraint that is not just finitely exact, or infinitely exact in the usual sense, it is transfinitely exact. This insight is very close to Anaxagoras' idea of chaos theory in ancient Greece. So there is a direct connection between symmetry and chaos and fractality, as we have seen before.

In Euclidean geometry, you can consider a sphere and this sphere has a symmetrical shape.

A spherical symmetry, yes.

On the other hand, we have some fractal shapes like the Mandelbrot set or Julia sets. I think they represent a new kind of symmetrical object. If you magnify them, then you find a symmetry that goes beyond Euclidean geometry. Do you think this is a new kind of symmetry in physics or geometry?

Yes, you can insist that self-similarity is a type of symmetry of its own, absolutely. This brings me to the concept of indistinguishability which likewise is a transfinitely exact physical concept if it is obeyed by nature, which it is. Pauli was the first to think deeply about this in the last hundred years. There is a special term which he coined—"exchange symmetry." It means that if you have two particles—like two eggs, for example—that are assumed to be really symmetric in the sense that each is identical to the other except for position in space (apart from mirror-inversion, perhaps), then these two symmetric objects obey laws that are not the same laws that apply when they are not absolutely identical in all properties. This was called "exchange symmetry" by Pauli. It means that when you exchange the two, you have changed nothing—absolutely nothing. That is, you have changed not just something that is infinitely small, in its differing across the two specimens, but you have changed nothing at all, absolutely nothing, zero. This is like religion if you so wish—the idea of a perfection that one would ordinarily not believe to exist in physics. This absolute symmetry now has predictive consequences that can be checked in nature. Low and behold, nature obeys the mathematical consequences of these absolutely symmetric laws.

Exchange symmetry implies, for example, the following: Assume you have two absolutely equal particles moving on a line like two solitons—solitons are waves that have the feature that they can pass right through each other undisturbed. If the two now pass right through each other, this does not mean that the left one comes passing through to the other side on the right, and vice versa, as common sense predicts. Rather it means that at the moment of passing through each other, the two specimens exchange their identities. The one formerly on the left returns to its own side, and so does the one on the right. The two now suddenly behave as if there existed a repulsive force between their centers that forces each to return—with an infinitely strong momentary acceleration. This is a phenomenon found by Pauli to be at work in chemistry and in physics as a whole. He called it exchange symmetry as mentioned. It is an important feature of quantum mechanics. People believe up to this day that it belongs to quantum mechanics, as one of its many new laws; for it was Pauli who discovered it and Pauli was an early quantum genius. So everyone thinks today that this is just something discrete—another quantization effect. They do not realize that this a feature which quantum mechanics shares with classical physics empirically, and that this phenomenon is transfinitely exact in its nature, not discrete. This means that it is as exact as only chaos theory or fractal theory allows for. Quantum mechanics for once is not a more or less discrete edifice once this effect is included. This effect is a classical continuous ultra-accurate—"transaccurate" would be a better-fitting term—phenomenon, evidence and proof of a transfinitely exact feature existing in physical nature.

I do not know whether this question makes sense if I ask it here; is nature symmetrical?

It is hard to talk about nature as a whole, but with Leibniz I would say, yes.

If I say "nature," I should perhaps clarify. Nature can be considered as a very big phenomenon, but we need models to explain or describe it. When we have a model, we consider physical laws. I can mention here a beautiful mathematical theory due to Emmy Noether. We know that, based on Noether's theorem, for every conservation law in nature there exists a symmetry, and for every symmetry there exists a conserved quantity. This is a very powerful theorem in mathematics and mathematical physics.

It is a very original idea and I love it, of course, it fits perfectly into our context. It simplifies a lot of phenomena and puts them under a common umbrella. Emmy Noether made a big progress; she was one of the most powerful mathematicians of the 20th century, working with the great David Hilbert. The three main discoveries she made in the area of physics are very intuitive: the translational symmetry, the rotational symmetry and the translational symmetry in time between past and future. They each explain a conservation law in physics. For example, conservation of energy is guaranteed if you assume that the laws of nature are invariant under a change of position in time. The second is momentum conservation which follows from translational symmetry in space. The third, angular momentum conservation, follows from rotation symmetry. It is as simple as that.

This is well known for almost a century.

Yes, and it happens that there also exists a fourth—new—law that follows from Noether's intuition and is very important in physics as well. It applies when you have a type of rotation symmetry that can be called a "layered rotation symmetry." On every level in a tower, say, you have rotation symmetry and hence conservation of angular momentum in the absence of friction. At the same time, the whole tower also harbors conservation of angular momentum across height levels. This holds true, despite the fact that the rotation rate of clocks is known to differ across height levels as follows from Einstein's clock slowdown downstairs. We recently talked about the example of a rotating bicycle wheel devoid of friction which was suspended from its hub and hauled up and down. This represents a new special case of Noether's theory. The wheel is rotating at different rates on different levels of gravity because it represents a clock and as such "ticks" more leisurely downstairs. Nevertheless the angular momentum of the wheel is conserved across heights. We can haul the wheel up from downstairs and go down again reversibly as long as we wish. The rotation rate changes but angular momentum is conserved. Therefore, one can prove that the size of the wheel and its mass or both is bound to have changed too because the conservation law of angular momentum is so strong that it actually forces both space and time to "give in," so to speak. Indeed size is being changed across heights (increased downstairs) and mass is being changed across heights (decreased downstairs), and charge follows mass due to their locally preserved ratio, a constant of nature. There are many further generalizations of Noether's theorem which can be found in this area, as György Darvas will predictably confirm in regard to his discovery of isospins.

Now I have a general question; do you think that information is conserved in the world?

This is difficult to answer. Something like information…

I know, the question is not very clear. I meant that if we assume conservation of information in quantum mechanics; what in this case is the corresponding symmetry?

The question is beautiful, but information is not well enough defined a concept, I am afraid.

I think information can be "created" but I am not sure whether this is true. Yet if we have conservation of information, then this would mean that we may also have a corresponding symmetry based on Noether's theorem?

Yes, but Emmy Noether insisted on continuous symmetries. Information is not well defined so far in this direction. If we look at fractals, for example, one cannot describe a fractal in terms of information theory. This is due to the transfinite exactitude of fractals. The amount of information in a fractal structure is transfinite, since if you see a certain little feature somewhere in the Mandelbrot set; it always has the same or closely related features no matter how small a specimen on the Mandelbrot robe you look at. Information in contrast has an intrinsic "discrete ring" to it. It therefore can only give an approximate description of a continuous curve, even though many nice laws hold true for information, too. Still the concept of information is not as deep as physics itself is, if you look at it from the point of view of dynamical systems and chaos theory. So "information" is a less robust phenomenon in light of the exactitude of transfinite mathematics. One can hope for a transfinite information theory eventually, but then you will not need the term "information" anymore, perhaps.

At the beginning you mentioned Leibniz' case of two universes. Let us talk about the idea of symmetry in quantum mechanics. Do you think we can find a correspondence between Everett's viewpoint in quantum mechanics and symmetry? Do we then have two worlds or just two simultaneous frames? Do you see any new or special symmetry here?

I never thought about the relationship between Everett and symmetry. The different Everett worlds would be different cuts through the same exo-universe. One way to interpret Everett's theory is that we have one exo-world, and within this exo-world, you have a cut and I have a cut, and each cut is different. The Archimedean example of frictionless heavy flat bowls moving on an icy plane was the first instance of a conservation law. If we assume such a frictionless billiard-balls type universe, as we can do, then it is a deterministic universe. If you want to look at it from the point of view of a particular subsystem existing inside, then a difference is applicable between that subsystem and the rest of the universe. Whenever the subsystem that you focus on is shifted, the rest of the universe will be changed relative to the subsystem in question. So there is a "difference principle" at work. This difference principle can be used to understand Everett's theory. Therefore, a whole new problem arises here: If the boundaries of the observing subsystem (organism) are redefined a little bit, then automatically the physics that the organism observes also changes. This is completely understandable in the terms we already dealt with in the spirit of Leibniz. Hence there can be no identity between the different Everett worlds across observers. I would not know how to apply the notion of "symmetry" here.

The conservation laws we have in physics can never be violated because they have a very good support by mathematical theories.

We observe that they cannot be violated and deduce from this fact that they must be transfinitely exact. We know this from indistinguishability, the Pauli phenomenon in chemistry. Indistinguishable particles do exist empirically even though this fact appears infinitely improbable at first sight. As soon as we have such a phenomenon, we know that nature is crafted with a transfinite accuracy. This transfinite accuracy is maximally intimidating. It is such a big momentous finding. All the discreteness and the probabilistic features of quantum mechanics pale before this infinitely powerful strength of Leibniz' and Noether's symmetry theory. It is such a challenge to have indistinguishable particles—to have two electrons—and to see the living proof before one's eyes that if they exchange their places, nothing has changed—not just infinitely little but nothing at all. This "absolute physics," as one could call it, is pervading quantum mechanics. So there is no doubt that this absolute theory is much more powerful than any more approximate or empiricism-based "direct" embracing of quantum mechanics can be. I believe in this idealistic thinking behind physics. The best proof is the finding of an absolute symmetry. Noether's theorems do not just apply because we have an idealized mathematics: this "idealized mathematics" makes tangible predictions that nature fulfills empirically. Therefore, nature herself is ideal, a fact which is not well known.

Let us consider an abstract universe in which we, for example, do not have conservation of energy or conservation of momentum, or in which some other conservation laws are violated. More specifically, let us for the moment suppose that we do not have symmetry in time. Because of this, we have no conservation of energy. What are the characteristics of such a world?

I talked about these things with my friend Bob Rosen many years ago. He had pointed out to me the "scandal" that Hamiltonian physics—and hence nature—is "structurally unstable" in the mathematical sense of René Thom. Take the harmonic oscillator, a very simple differential equation, a frictionless pendulum, as an example. Here physics shows this remarkable time reversibility of Hamiltonian systems which is structurally unstable. An infinitesimal perturbation in the mathematics (a non-zero damping) destroys Hamiltonicity, destroys frictionlessness. So Nature is under the baton of a very, very delicate angel as it were, if conservation laws make sure that structural instability holds true in spite of its infinite delicacy. Observation-wise, structural instability amounts to an infinitely strong vulnerability. So nature is infinitely weak if you so wish, but on this infinite weakness we can absolutely rely despite the fact that this reliance means relying on something that is infinitely unlikely. We are confronted here with a feebleness much more delicate than a spider's web. Thus nature is infinitely sensitive, infinitively delicate. This delicacy of nature was forgotten for a long time because people had grown accustomed to the discrete features of quantum mechanics, which look so robust, so insensitive. This pseudo-discreteness follows from the fact that quantum mechanics can be deterministically deduced as a property of the interface between a subsystem—the observer—and the rest of the world. But why can we not see that interface? The answer would be that this is because we are infinitely accurately confined and bound to some of these infinitely accurately described elementary particles that inhabit our brain at a given infinitely accurate moment in time. In the brain, there would be a sub-portion—some cells or perhaps a single cell or a part of a single cell—at a given moment that harbors consciousness in the sense that consciousness is attached and confined to it. This subportion of the physical world to which our consciousness is tied at a given moment would then divide the world in such a way that we are confronted with quantum mechanics. The difference is the world as it is described by quantum mechanics. All the measurements that have been performed in the past and, for example, are printed in the books along with their mean values (which yield the value of h-bar in the books) would be described by the Sackur-Tetrode equation valid for this momentarily "switched-on" subsystem. They would all be subjective measurements, results that are true only relative to one of these strangely and infinitely exactly specified subsystems. I apologize for this proposal. It almost imposes as a religious type of thinking on physics. Infinity is a theological notion if you so wish, and so is symmetry.

What do you think is the relationship between symmetry and art? There are many beautiful buildings in the world, erected centuries ago, which often display a very nice symmetry, stemming from different cultures in different areas and on different continents, but the common thing is symmetry. What is the relationship between symmetry and art? Is this sort of symmetry different from the symmetry found in physics?

The mathematics is the same. The human mind understands that there lies a challenge here of a metaphysical character. Symmetry and metaphysics belong together, and art and metaphysics have a mutual affinity, too. It is a little bit of a pity if some religions, including Judaism and Protestantism, were somehow afraid of pieces of art. Maybe it is a sign of respect, but art and religion and symmetry are three noble fields that are closely bound together since the era of the Shamans.

In paintings and in buildings we encounter symmetry; is there also a relationship between symmetry in physics on the one hand and symmetry in music on the other?

Johann Sebastian Bach made some transpositions of some melodies and you do not recognize them afterwards. Temporal symmetries are especially hard to fathom by the mind. Susi Vrobel has a book here, called "Fractal Time." Some minds are able to understand this better than others, and I am afraid I would not be up to saying anything perceptive in this regard. A friend of mine who is a professor of music and at home in ancient Chinese thinking, Anthony Moore, would be a person to ask this question.

We already talked about the symmetries present in fractal geometry. What is the meaning of symmetry in chaotic dynamical systems or chaotic attractors?

Here I would mention Christophe Letellier who could answer this question much better than I can.

But if I ask you to answer?

I try to put myself into his mind as best I can. He loves symmetric chaotic pictures and he found a way to multiply an attractor like a spiral-type attractor into a multi-faceted, very symmetric structure. There are also beautiful theorems on that, but again, I withhold judgment. I am not good enough for this type of question I am afraid.

Can you recognize a new Noether-type conservation law in chaotic attractors or not? In chaotic attractors, the symmetry is a kind of geometric symmetry, is it not?

It is an unusual symmetry. I mean, take the Lorenz attractor, the so-called butterfly attractor of Ed Lorenz. It has a strong symmetry to it, but you can reduce this symmetry to make a less symmetric object out of it. So symmetry is an available option in these mathematical dynamical systems. You can play with it, but the essence of chaos is this transfinite exactitude of not just every hair being counted on the head, but of a transfinite exactness of the counting as it were. Mathematically—for example in the Smale solenoid attractor—we encounter self-similarity with its transfinite accuracy. We already talked about the fact that the attractor in the Smale solenoid, even though of zero measure at the end, is invertible such that every point of the original volume of the extended basin of attraction that is drained by the attractor is still represented on the zero-measure attractor itself, and so in a reversible fashion if we neglect a set of zero measure. This transfinitely exact invertibility was already Anaxagoras' big insight. There exists a still unresolved difference of opinions about this. Steven Smale has never contradicted my transfinitely oriented interpretation—although I have to admit that I never saw him again after our two creative encounters in 1971 and 1976. From György Targonski who held a mathematics chair in Marburg I learned the secret of the "red line" (as I had dared to call the attractor in a talk in allusion to a fairy tale of the Grimm brothers in which a head had first been cut off and then glued back successfully, with this fine vestige remaining as the only trace). He accepted this term for the attractor of a dissipative chaotic system, but there may be other mathematicians like Smale who have yet to be convinced.

A last question; do you think that we can have further conservation laws in nature due to the discovery of new symmetries?

Yes, the example of the wheel that is being hauled-up and lowered reversibly referred to a symmetry that had not been described before. It is a symmetry in which angular momentum is preserved even though rotation rate is not preserved. One could use this example as a heuristic recipe to find other nontrivial types of symmetry which have not yet been described. There also is György Darvas, my other friend György, in particle physics who has beautiful theories in this direction in connection with isospin conservation as I already mentioned. The CERN experiment in elementary particle physics covers many sub-nuclear symmetries which deserve to be studied better in this context. High energy physics is one of the areas that depend crucially on the theory of symmetry, and in which new symmetries can be investigated. So your question makes much more sense than I can meaningfully answer at the moment.

Thank you.

Chapter 13
Language and Smile: Benevolence Theory

In this session we want to consider the subject of human language. There are different viewpoints on the evolution of language held by philosophers, but I want to look at human languages from a dynamical-systems point of view. Before concentrating on language as a dynamical system, I would like to ask you; how do you look at the history and evolution of language? What is the difference between human communication on the one hand and that of other animals or life forms on the other?

The history of human languages is apparently much older than people tend to think. Recently I saw a picture of a stone instrument called a "biface" or hand axe, a stone instrument used 10,000 years ago, but that specimen was more than one million years old. At that very early time, human ancestors had already developed the civilization to produce very finely carved stone axes. The use of fire is even two million years old. I have a strong opinion about the difference between human language—"languaging" as Andy Hilgartner calls it—and animal languaging. It has to do with an older theory of mine that personhood—being a person—is something specifiable scientifically, not just in the sense of one's being an individual but as representing an altered functioning of the brain, the function change in the sense of Robert Rosen as a hysteresis-type phenomenon. When one has steam heating in one's flat, and you change the temperature by turning the knob on the heater, then if you turn it back to the original reading, you will not get the same heating level as before because there is this "hysteresis" involved with most models. In other words, the same setting of parameters can have the system in two states—either in a low or in a high heating state as in the present case. This Bob Rosen called the "principle of function change", the fact that in the same system subjected to the same parameters or external conditions, a different dynamical steady state or attractor can nonetheless be at work.

In the same vein, if you have a brain, the latter can function in different modes under the same external conditions. The same brain with the same experiences and hormones etcetera can function in different modes under equal input conditions. It may therefore be possible to completely upset the functioning of a brain so that it

A. Sanayei and O. E. Rössler, *Chaotic Harmony*,
Emergence, Complexity and Computation 11, DOI: 10.1007/978-3-319-06781-0_13,

functions in a radically new way compared to before, under the very same condition. One knows about these things hypothetically in the context of psychiatric diseases where physicians indeed think that some psychiatric conditions are based, not on a change in hardware but only on a change in looking at the world through the software if you so wish. So when one agrees that the brain can support many types of functioning including even qualitatively different ones, it makes sense to wonder whether the difference between animals and their "languaging" and humans and their use of language is not an example in the same category. If this hypothesis is correct regarding the human brain, the brains of some animals might be susceptible to the same bifurcation phenomenon. That is, if the human brain is making a different use of its own capabilities, then also some animal brains could predictably be used in an analogously different manner, and conversely also the human brain can be used in the animal way in principle. To me, the fact that human beings are "persons" and thereby differ functionally from animals is not necessarily due to the fact that they have different brains but that they have "snapped" into a different usage. Then what remains to be explained is how come that human beings make a radically different use of their brains. This is the subject matter of a theory of the human smile. Jan van Hooff in the Netherlands wrote a big paper in 1972 titled "A comparative approach to the phylogeny of laughter and smile," comparing smile and laughter in apes and humans. You will be surprised that I at first talked about states of dynamical systems including brains and now suddenly talk about smile and laughter which appears incongruent at first sight.

We talked about the brain equation before. How can you explain the ability of human beings to use language in light of your brain equation? What is language actually?

Language in the human sense presupposes that the speakers are persons. And to me, being a person represents a different dynamical state than being not a person. The brain can support both being a person functionally, and being not a person functionally. This seems to be a very broad distinction. If true it need not be confined to the human species. Then a different type of functioning can be predicted for many brains which share certain characteristics.

Would this person then be human or not?

In one sense yes in the other, no. You can take an animal brain that is mirror competent. This is the one condition. Human children are mirror competent from a very young age on—it is heart-warming to watch the joy of a toddler when presented with a mirror. Many apes are also mirror competent, apes and elephants and some other animals like orca whales and dolphins, also magpies and keas and the impressive Newcaledonian crow. Perhaps also some invertebrates, like the giant octopus or some mantis shrimp species join the same fold—these animals with the most sophisticated eyes on the planet. Most of the latter species have never been studied even superficially. "Mirror competence" means that you can use the spatio-temporal simulator built into the brain (I call it the "great simulator" placing it beside the "great motivator") in such a way that while you observe something, you can simultaneously put yourself into the position of what you are observing. In neuro science, not long ago the so-called "mirror neurons" were discovered by Giacomo Rizzolatti. Most higher animals possess mirror neurons, but mirror neurons do not yet guarantee mirror competence.

Mirror competence was first described in 1921 by Wolfgang Köhler, an elder friend of Konrad Lorenz. He was an ape scientist and creatively observed his chimpanzees, young child-chimpanzees, in Teneriffa in a gestalt-psychological setting. He made important experiments with these apes, very benevolent merely observational experiments. In a famous experiment he put different-sized boxes into the cage and had hung a banana from the ceiling before the apes were let in. The animals were supposed to figure out a way to get at the banana. Frustrated at first and for quite a while for not succeeding directly in several ways, they used their intelligence so as to suddenly realize that you could take these boxes and put one upon the other and then climb up. The sudden insight was the remarkable thing. What they never learned was only to first take the big box and then the next-smaller one and so forth. They were just running up the random pile of boxes to catch the banana while underneath everything would collapse. But they had gotten the banana. The "aha!"-effect, the sudden transition from initial despair to the joyful anticipation of the success was a major discovery in what was called "gestalt psychology."

To come back to our context of mirror competence; the same Wolfgang Köhler also observed another behavior that the same animals exhibited. They were held in a cage that became quite cold in the morning and their urine—forgive my mentioning this—had become frozen on the floor during the night. They discovered that they could use the flat slippery surface as a mirror, to look at their own image. Then Köhler gave them pocket mirrors to play with and was not astonished to see that his apes would look at their own teeth and other hidden parts of their body in this way. So they turned out to be completely mirror competent. This mirror competence is a type of functioning of the brain which presupposes that the built-in simulator is very highly developed. You then cannot just anticipate the correlated changes that occur in the environment as you are moving through it, but can even recognize yourself in a mirror because while you are doing something here you are aware that the same activity of your hand or whatever, is showing on the other side in a different orientation. To identify oneself as being reflected in a mirror is a very high-ranking performance and capability of brains. Only the animal groups that I already mentioned are able to do this, but if you have this capability of putting yourself into the shoes of the other side, you can exploit this fact, and these animals do this, of course. This feat is not enough to become a person, even though most everyone would naively think that it does. One has to ask oneself, Why are these animals, these apes, that can recognize themselves in a mirror like a small child, not also persons like human children? Everybody thinks that this is because they are too stupid for that, but this is not true: They have no reason to become a person. I mean, the way we use language presupposes that we are persons. What does it actually mean to be a person? It is hard to define what a person is, but it has to do with the smile.

Here I have a question. For people who are praying, is there a communication between a person and someone who is not a person?

I am not sure that the addressee of a prayer is not a person. Prayer is a very difficult topic. Becoming a person also implies thinking of a person that stands behind the world. Praying is in effect, in my view, talking to a person even though it is an imagined person. Young children become persons around the age of 1½ or 2 years or so. There are no studies, to the best of my knowledge, attempting to record this transition of pre-personhood towards personhood in a young child. It is predictably sudden, and a very moving situation at that, which fact explains why there exists no recorded evidence in science so far. The idea of benevolence perceived stands behind personhood. A young child suddenly acquires the suspicion that Mom or Pa wants him to enjoy something, to love this fruit or to be happy. The suspicion of someone wanting you to be happy, becoming possible in the presence of mirror competence plus a factor X, lies at the root of one's becoming a person. How do I come to say this? It is because of the phenomenology of the smile and the laughter as Jan van Hooff already described it. In apes including humans, there is this evolutionary observation that the smile and laughter have eventually converged in one particular species, namely the human one. In humans, smile and laughter look virtually the same. What does a smile mean? A smile represents an innate behavioral pattern signaling that you are happy. But the smile also means that you are friendly, that you are bonding. There is bonding friendship between the person who smiles and the individual who is smiled at. The smile is an expression of bonding and laughter is an expression of happiness, of exuberant joy, but laughter has nothing to do with friendliness directly. Sometimes it can happen that if a whole group is laughing and you are passing by, you wonder whether they are not laughing at you or about you. So laughing is not necessarily something friendly, it is just an expression of joy. The smile is an expression of bonding. Nonetheless these two different expressions, which are completely disjoined in chimpanzees, have converged in the human species 2 million years ago. This convergence between smile and laughter characterizes the human species biologically. If you are treated in a friendly way by your mother as a young child and you see her smile and you laugh because you are happy, then you reward the mother by your own being happy. This is also the case in many animal species where the expression of joy shown by the young rewards the feeding adult. But the human mother will then, by being made happy, also laugh. The laughter of the mother now looks like a smile to the child. Therefore the human child unlike the young of other species is rewarded by the mother being happy. A transposed world. Hence both are being rewarded by the other being happy; the mother because it is normal in child-rearing animals that the happiness of the young is a rewarding signal for the feeding adult. We also have the opposite, that the happiness of the mother—because it looks like a smile even though it is just laughter, is a reward to the child. So the child predictably starts parenting the mother. Both animals, if you look at them as an animal species, are rewarded by the happiness of the other: "Cross parenting." This is a dynamically unstable situation, the basis of a positive feedback. It observationally gives rise to a "bonding bout" as it is called. Both grow into the rituals of happy bonding much more frequently than if the reward was unilateral as normal. This is not because they are both parenting, it is because they are both bonding even though only one is parenting, but it is nevertheless symmetric. This symmetry—again a symmetry!—is pathological in the biological sense because it only makes functional sense that the parents are rewarded by the success of their feeding the young, so that the latter survive. Now, the young are starting to feed the adults! This is the distinguishing feature of the human species in the eye of a biologist descended from Mars—that in these terrestrial animals, the young are feeding the adults. This occurs nowhere else in the biological kingdom and presumably in the galaxy.

Is it special for humans?

Exactly, it is the distinguishing functional feature of humans. If you come from Mars and are a biologist, then you will realize that this is absolutely unexpected.

That a child tries to put food into the mouth of a mother?

And asks: Is it good? The child wants to make the mother feel good. That is pathological in an evolutionary sense. You therefore have to classify humans biologically as the "Pongo goneotrophicus." In zoology, "Pongo" means great ape and "goneotrophicus" means parent-feeding. From a biological point of view, humans are the parent feeding apes. This is unusual in biology and you could consider it as a "lethal factor" in the sense of evolution theory. One can explain how this evolutionary accident occurred. The convergence of laughter and smile is the reason. It has to do with the re-acquisition of the pair-bonding drive at an earlier evolutionary stage in human pre-history. There is an apparent counter argument to what I just claimed. The human species stands not quite alone with bonding and happiness looking the same. There is one other species—namely the wolf—in which this also happened, although in a less symmetric form, with the familiar tail-wagging of the dog and its ancestors. Hence if this condition alone were sufficient, the wolves would also have to become persons, because again the young could misunderstand the excited happiness of an adult as a rewarding signal that causes the young to bring a sacrifice in effect. Whether or not this can be observed I do not know, it was never looked for empirically, I think. The main consequence which arises in the human species would nevertheless not happen in the wolf or dog. The reason is that the wolf is not mirror competent. They therefore cannot put themselves into the position-in-space of the other side. Only if you combine the trait of mutual rewardability by the happiness of the other with mirror competence does there arise a second interactional instability—the decisive one which leads to a function change epigenetically. This second instability completely changes the functioning of the brain of the human child. Just as it did with the mother when she was a child. If the mother had not been a person before, the mother would likewise become a person in this interaction. This is what happens:

When the mother is laughing because the child is happy, the mother is rewarding the child in addition to its already being happy by making him more happy, since rewards are additive. Since the child is able to put himself into the shoes of the other side by virtue of his mirror competence, he can and will on the first or the tenth time acquire a strange suspicion; Mom wants me to be happy. This "suspicion of benevolence" then gives rise to an experiment that the child initiates, a test; does Mom really want this wonderful happy experience to occur? It is a suspicion about an intention that is not real, not being held here but somewhere else, and it is benevolent: strange suspicion indeed. It arises subtly and tentatively at first. The toddler is, for example, exaggerating his happiness, his anticipation of being fed by opening the mouth unnaturally wide, to see whether this works. Predictably, it does work and even gives rise to a common laughing bout. Mom is even more rewarded, becomes even more friendly and so, after days or weeks, a full-fledged bonding bout develops out of being fed. The less pronounced parent-feeding tenderness in the wolf does not lead to the moving bonding bout observable in humans which introspectively is accompanied by the suspicion of benevolence existing over there, in a foreign will. So the idea of benevolence—the invented absurd concept of benevolence—is what distinguishes human beings, human persons, from other animals. The suspicion of benevolence is like a fire, an interactional instability which completely transforms the functioning of the two brains involved. It is characterized by many active tests made and jokeful interactions and an inexhaustible creativity in making "fun" together, a maximally humorous and moving event at the same time.

In Artificial Intelligence based on a biology-related model, namely the brain equation, you will predictably have the same phenomenon, an artificial person arising out of a computerized hardware. The essence in either case is the invention of the suspicion of benevolence. The "animal brain"—a formally autistic brain since all brains are autistic by design—becomes suddenly a person. That is, it employs cross-anticipation in a mutually rewarding fashion. The child becomes benevolent himself; he wants Mom to be happy, too or at least Mom to be happy, even at his own cost. Major—stupid-appearing—sacrifices have been observed to be brought by young children. The essence is that it is all too nice to make a game out of realizing that one is able to make the other happy, and to be able to absolutely rely on this. And to test this back and forth in jokes. Both are thinking or realizing that they are infinitely valuable for the other. This is called "love," but it is not love in the sense of attraction, as society believes, it is love in the sense of wanting to make happy at all cost. It is holy.

How do you explain the procedure of learning a language? When a child tries hard to learn a language from his parents, is it part of the same process that you explained in benevolence theory?

Language is much less sophisticated than the idea of benevolence. You can have language without this transformation that I described. You can teach animals to understand language in the form of announcements or orders very easily. Horses understand what to do—understanding of an order is easy to get using plain reinforcements. Learning to give orders is also something animals that are not persons readily acquire. Language has these two sides to it, on the one hand it is just a communication system which can be used and understood and is made use of by everyone. That is not real language—you could call it imperative language or training-to-perform language. Humans at a very young age often use language passively and actively in this restrictive sense of understanding orders or giving orders. The use of language without personal pronouns or with a totally false use is compatible with the suspicion that this language is not yet a personalized language, provided it only consists in following orders or getting effects by uttering certain expressions. That can also be called language, but it is not personal language even when the grammar is not underdeveloped. You could call it "pre-language." This pre-language is something that many animals can learn. Some birds, intelligent parrots, for example, have learned it. The personal use of language has not been observed to my knowledge so far in a non-human species, but it can predictably be evoked. It is a grave responsibility that will thereby be incurred. Do I have the right to mention it? It is no doubt a moral duty to try this in the case of smile-blind human children. The fact that smile-blind human children are not the only smile-blind intelligences that are eligible—and have the holy right to be helped—then follows as a corollary. The method to follow is straightforward; whenever you are happy, you have to reward the beloved mirror-competent bonding smile-blind partner, like a young chimpanzee or orangutan or elephant. You only need give the right bonding signal via a channel that works (for example, the acoustical one) whenever you are acutely happy yourself in the interaction. A so-called "acoustic smile" in the case of a smile-blind child will predictably work. The right acoustic or infra-acoustic or gestural sign in the case of the young white elephant whom you want to rescue. We need no longer keep up the false belief that there exist no more highly developed brains on the planet than the human one. The first holder of a patent for the atomic bomb, Leo Szilard, even wrote a fairy tale after the bomb had been dropped, "The Voice of the Dolphins," in the desperate hope to let intelligence rain onto the planet.

Are you serious with this?

I first published the basic idea in 1968, and the trans-specific option in 1975. Only Gregory Bateson whom I visited in the same year understood it at the time. Konrad Lorenz said it was "too difficult to understand" for him. The idea is related to Skinner's operant conditioning, but nevertheless totally different. A positive feedback develops when you feed the dolphin and you reward him in addition by your own joy, displayed as bonding. The mutual reward will develop in the same way as it did between a human mother and her smile-blind child, when she softly acoustically expresses her joy at a wonderful feat accomplished by the child sitting on her lap learning to write. This situation occurred spontaneously in a documented case (which I by chance once witnessed in a Stern-TV report, on January 28, 2008). Both suddenly develop the suspicion that the other wants her or him to be happy. To invent the suspicion of benevolence is pathological in the animal kingdom. We saw that it represents a misunderstanding originally based on an evolutionary accident, convergence of smile and laughter, in one particular species. At the same time it is a "jump up towards point Omega" in the parlance of Teilhard de Chardin. It anticipates in the form of a jump the asymptotic endpoint of evolution in the cosmos. Love has this miraculous nature. It is a self-fulfilling prophecy. Even if the actors realize that it is a misunderstanding, they can check on this, and when they do, it turns out to be real. The child is at first testing how good, how strong, is the love of the mother and whether it is really true that his joy is desired over there. So the whole thing is not just a misunderstanding, it is a misunderstanding that can be checked and can be found out to be the truth. If it suddenly proves confirmed, they both recognize the other as a person who wants the first to be happy as a different person. With the scientific understanding of the unbounded crazy love of the child comes also the understanding of why we as persons are able to love each other. The ultimate reason is, if I am allowed to add this, that within this positive feedback that we just touched with our minds, a third person is invisibly involved. For the players did not make themselves and get the rewards of color and light and the Now from a third side that turns out to lie beyond the realm of science. So, genuine love—in the sense of the miracle of personhood arising in the young child—is automatically connected with the miracle of color and the Now as a present given to both.

How do you explain grammar in language? If you, for example, consider the German language, you find very special rules, some not admitting any exceptions. If a foreigner has to learn a language, he has to be pointed to the special rules in that grammar. How has this grammar originally been created? Did the earlier humans invent some rules for their talking?

Sometimes people of different languages are forced by circumstances to acutely live together and to suddenly have to communicate somehow. Then they develop a sign system. By using redundancy in communicating repetitively, the redundancy becomes a rule for both. The capability to make this joint invention is automatically implicit in the natural benevolent relationship between persons. So the grammar is not really a mysterious thing. It is an automatic implication of persons agreeing to communicate. It is like a straw, something that is lying around ready to be used as a symbol for anything. It is much easier to communicate along roads that have already been used—like elephants in a forest tending to use the trodden paths. You would not call it a language if animals or ants follow the same path, but the principle is the same in both cases. So "grammar" is not as high-ranking a feature as linguists sometimes tend to think—notwithstanding its lawfully changing awe-inspiring regularities as the families and superfamilies evolved over time.

A last point; natural languages can be juxtaposed to computer programming languages. Lotfi Zadeh, the inventor of "fuzzy logic," says that the human languages obey a non-binary logic which he calls fuzzy logic. He claims that we can deduce, we can measure, we can compute with our natural language without a lot of information or sometimes any information. For example, when you want to cross a street, you see some cars coming towards you but you do not have any exact information about their speeds, and you also do not have any exact information about the friction coefficient between your shoes and the ground, or the speed of the wind. But by an approximate checking, whether the approaching car's speed is high or low, you can make a very reasonable decision in this inexact way, based on your previous experience. Do you agree with the idea that natural languages use another logical system, namely fuzzy logic, in contradistinction to the strict bivalent logic?

I also think that the bivalent logic is something very artificial as a late discovery of humanity. It is ultimately also implicit in the richer logical structure of a language. To steer one's way through a jungle is at least as difficult as to steer a motorcycle through a traffic jam. This also is something that not only humans can do, but many animals are apt at doing—without bivalent logic. It is implicit in using the brain in an intelligent fashion employing spatial thinking.

Let us look at the essence of this intelligence. It seems to use a special logic for its computations?

If you want to use the word logic for the description of any dynamics, then you can use the word, but essentially this is just optimizing one's way through space. Any autonomous optimizer is doing exactly that. The brain is such an optimizer, and we have this time-dependent optimality functional that controls our behavior. The latter is implied by the brain equation. Only if we are conditioned by grammar will we suddenly begin to wonder how this analog-type thinking enters into our fixed language system. It is the other way round, the former is natural and language is something which was later formed as a very strict grid of rules that are, if you so wish, discrete. The underlying analogical thinking is not discrete. This non-discreteness we can wonderfully approximate by the notions of Lotfi Zadeh, which form a powerful method. It is a strictly ordered type of thinking that language supports—language is very logical even in a discrete sense. The burden of the discretization of thinking, implicit in language, is being alleviated by the new fuzzy logic. We are presently on our way back towards a new analogical thinking, with the iPad, for example.

Do you see any relationship between the dynamics of a language and chaos theory? Can we have a chaotic model for the use of language?

The brain equation itself is chaotic in the behavior which it generates. Very small changes in the initial conditions lead to divergent trajectories. So chaos theory is automatically applicable to the autonomous simulator in the physico-mathematical machine called the brain. And the brain is also a chaotic machine, since it is a nonlinear equation that is implemented. Thus, I see no problem here, but your question was more poignant.

Is it possible to have a model for the whole of a language as a dynamical system or a differential equation, or is this not possible?

You can model the whole brain as a differential equation. The brain equation is a case in point.

As an ordinary differential equation?

I would even go this far.

Then you consider your parameters as exact parameters, or can they vary in an interval?

The brain equation would be an exact dynamical system, but its behavior will show these Zadeh-type tolerances. So the optimizer is fuzzy in effect. The optimality criterion needs to be satisfied, but there is always a set of possibilities to choose from, all smoothly continuous. This fuzziness is part of the nature of the brain equation, and only language makes an exception by being discrete and digital in many respects.

If you consider your brain equation as a fuzzy differential equation, in the sense that your coefficients can vary to a certain degree, would you think that it becomes more powerful in this way?

I would tend to think that a differential equation is fuzzy by definition in the sense that what it is producing depends on some momentarily time-dependent parameters, but that it is not fuzzy in its structure. It is a dynamical system which produces so-called "qualitative behavior" in the sense of Poincaré. This qualitative behavior is fuzzy in many respects; it is almost another word for fuzziness. But you would not need to add fuzziness to this fuzzy system because the deterministic dynamics produces a qualitative outcome anyhow. Qualitative dynamics and fuzziness are parallel notions. "Fuzziness" could then be seen as a way to re-discover this intrinsic qualitative structure in a context in which we previously only had discrete tools available. So a discrete system plus fuzziness added would be able to closely simulate the original qualitative continuous deterministic dynamics.

Would this explain why Zadeh titled one of his papers in 1996, "Fuzzy logic = computing with words?"

This looks reasonable, but the words here again have a meaning which is qualitative. So with respect to words it is correct to say that language is discrete, but the meaning of the words is qualitative. Therefore there is a residue of qualitative dynamics left within language and you can call this "fuzziness" again. There exists a close relationship between the fuzziness of a discrete set and the continuous qualitative transformability—a topological feature—of deterministic dynamical systems in chaos theory, for example. So fuzziness forms a bridge between the discrete type and the topological type of thinking. Poincaré was a person with a strong mind in both fields. Topology, Poincaré's brainchild, is related. Lotfi Zadeh is an ally to Poincaré in this respect.

Do you have a further complementary idea?

In addition to the above? I had not known before the things that I just told you. So I am grateful that I was allowed to develop them in interaction.

Chapter 14
Three-Body Problem, Poincaré Recurrence, Homoclinic Points

In this session we will change the subject and come back to more fundamental questions in physics. The main subject I propose to talk about is the many-body problem. As a first step, let me ask you; what is a two-body system? What are the special points? And if you like, we can then add a further body and discuss the phenomena characteristic of three-body systems?

There is a certain risk that we eventually touch the topic of chaos. The 2-body problem was seen by Kepler first, formerly a student here in Tübingen. He found this beautiful law that orbits can be elliptical, and in addition that if an elliptical orbit is followed by a planet, the area between two rays drawn from the sun to the planet in equal time intervals is the same no matter how far or close the body is to the sun. If the body is very close, the angle is large because the body is moving faster closer to the sun and if it is far away, the angle is smaller—a beautiful simple law. The discovery that two bodies followed such a both nontrivial and simple law was very surprising. Newton would not have been able to come up with his wonderful equations without Kepler having found this fundamental result. It is also surprising in retrospect that if we have two bodies, they can only follow a periodic motion in their dance and that this motion has one of the two focal points of an ellipse as its organizing center. There were clearly more options, a priori speaking, than following a flat elliptical orbit. In general relativity, indeed the simple ellipse is lost and you get a rosette formation in the form of a slowly circling open ellipse, a quasi-periodic motion. People have worked a lot on these problems for 3½ centuries since Newton first saw the solution with his correctly divined equations. Hill later found the reduced 3-body problem. We now already talk about three bodies. As soon as you have a third body interacting with the two, planet and sun, like the moon, everything becomes "messy." The moon is too small to have much influence, but in other situations three bodies suffice to generate a chaotic tangle. I never looked at this closest 3-body problem in detail, but it has been known for centuries that as soon as you have three bodies, you can no longer write down the solutions in contrast to what Newton following Kepler could do for the 2-body problem, although Newton's equations remain valid.

A. Sanayei and O. E. Rössler, *Chaotic Harmony*,
Emergence, Complexity and Computation 11, DOI: 10.1007/978-3-319-06781-0_14,

There is no analytical solution?

One does not have a complete set of integrals to describe what happens. Thus one has to use approximate methods. Poincaré tried to write and indeed wrote a long mémoir about the stability of the solar system in response to a question posed by the Swedish academy. It got accepted by this forerunner to the Nobel foundation. Then he found out that he had made a big formal mistake. As mentioned, he bought back the whole edition of the book that had already been printed, from his own money to get the error eliminated. For the 3-body problem had suddenly taken on a life of its own in his mind. We know today that this was the origin of chaos theory. Poincaré was the first to see the infinite complexity of the tangle of trajectories in space formed by the 3-body problem. He even said that the mind is turning away in horror from the complexity that has arisen in the 3-body problem. I always planned to check how Poincaré found the method of the Poincaré section which enabled him to see this, a so-called 2-dimensional map.

What was the mistake in the first special work?

The mistake was that he had overlooked the existence of chaos. So by writing the big mémoire with the error contained in it, he had become strong and exacting enough in his work to break through to a new clear picture. Within this clarity, suddenly the complexity of chaos took shape. No one before him had achieved such a depth in the 3-body problem and in mathematics. He found a way to identify the simplest case with three bodies which gave him a clue as to what happens—what is the source of the complexity. I know this only from Igor Gumowski who had a great influence on me. He told me that the reduced 3-body problem that Poincaré discovered consisted essentially of a very special configuration which was physically artificial but mathematically prototypical, a typical thought experiment. If you have one big body, which could be the sun, the sun can for simplicity's sake be assumed to have a permanent hollow tunnel running through its middle. Within the latter the second body can then move up and down periodically which is an allowed version of the Kepler problem with the ellipse having become infinitely thin.

Is this similar to the "T-tube system" which you already explained?

This is correct. It is also closely related to Newton's first intuition. He allegedly saw an apple fall from a tree during this 1 year of forced absence from his regular studies because of the plague. It is quite plausible that he indeed visualized the Alice-in-Wonderland experience—that there was a hole opening underneath the apple through the earth. And that in this tunnel, the apple would be able to fall and fall right through the middle of the earth and upwards again on the other side—until it predictably would reach exactly the same height above the ground again which it had had when it lost hold of its tree, touching the leaves of an imagined "anti-tree" in New Zealand on the other side of the earth. Then after having made this motion, the apple would return to come back up to the original branch on the original tree from which it had fallen, assuming there was no air resistance in the tunnel.

This day-dreaming thought which occurred to Newton showed him that what he had looked at in his mind was a degenerate Kepler ellipse. If the apple had had a very slight push to the side by the wind while losing its grip, it would have fallen in a very narrow ellipse. As soon as the idea of an ellipse had occurred to him for the falling apple, it was clear that the ellipse could be widened further. There exists a drawing by the hand of Newton which shows a tower on the surface of the round earth, and a stone falling first very close to the bottom of the tower, then—if given a horizontal push—it follows a wider parabola or upper part of an ellipse, and as the initial horizontal push is made stronger and stronger, the ellipse followed by the ball is big enough to no longer quite touch the surface of the earth. Then it suddenly follows a closed path just as Kepler had found. This day dream of stones falling from Galileo's tower after having been given an initial kick to reach a circular orbit around the earth was apparently the original intuition which allowed Newton to eventually write down his equations describing the orbits—the Hamiltonian orbits of planets.

And Poincaré?

Yes, this very intuition of Newton's was resuscitated by Poincaré who now needed a third body. He had already these two bodies of Newton's, the solar sphere with a vertical tunnel in the middle and an up and down oscillating planet inside. He could then choose for the third body a moon circling around the equator of the sun. Now there were three bodies, one fixed (sun), one going up and down (earth), one circling the fixed one (moon). The vertically moving earth would perturb the circling moon but much less (almost not) vice versa. What will these three bodies do in terms of their interaction? Hereby the big body (sun) can be made big enough so one can neglect its influence altogether because it is constant. Then only the interaction between the horizontal moon and the vertically moving earth needs to be considered. We understand what is bound to happen. When the vertically moving earth is close to the middle, the circling moon suffers the strongest attraction. We can simplify further by putting the circling moon into a frictionless glass tube so only its speed but not its radius can be affected. Then the vertically oscillating earth will, as it approaches the middle of the sun, more and more strongly attract the circling moon whereby the velocity of the latter is bound to increase. The converse holds true when the earth is moving away from the middle. Thus one can start to completely understand this maximally simplified 3-body situation.

The next step in our trying to decipher how Poincaré's mind worked is to no longer look continually at what happens in this system—a task that is still hopeless—but rather to make sure one looks only at well-defined recurrent situations, such as when the circling moon passes right before the middle of the sun (cuts a vertical half plane). We can at that moment record two variables, one being the momentary speed of the circling moon as it crosses that plane, the other the momentary positive or negative speed of the vertically moving earth. These two numbers define a point in a plane. Over the course of time, the successively obtained points lawfully sprinkle the obtained "Poincaré map," as the picture is called. Other choices of observables are possible, giving related results.

Whenever it is a three-body system, the map is two-dimensional?

By no means. The assumed symmetries were necessary to obtain a maximally simplified but still understandable case which then turned out to be infinitely complex. Poincaré's loving patience is perplexing. Note that the system in question has at least 17 variables (three positions, three momenta in every space dimension, minus one due to energy conservation). To end up on a 2-D plane that still captures the essence is like a miracle.

We have a lot of different options how to choose this map for the tangle of trajectories?

Yes. The mentioned particular map is convenient for interpretation. We know the horizontally circling body is at the same angular position again, we can record its speed, and the position of the vertically moving body at that moment. We find that if the vertical body is in the middle position, the speed of the horizontal one is at a maximum. If it is farther away from the middle, the speed of the circling body is smaller. So what we get is a series of points. If one looks at the resulting map, one eventually sees that these points each mark the edge points of little rectangles in this 2-D plane which have the same area. This is an area preserving map as it is called. What we then see is a so-called quasiperiodic motion. Everything is absolutely transparent. The quasiperiodicity means that eventually all points lie on a continuous connecting line in this 2-D plane, a line which is going to be filled-in in the limit. There is no chaos so far, but then, what happens if we change something? For example, we may perturb the circular orbit of the horizontally moving moon in its glass tube into a slightly compressed form, an ellipse, or further into a very narrow ellipse. The points that formerly lay on a smooth one-dimensional line in the map then no longer do so. There is more structure now. This is the point at which I cannot say how Poincaré proceeded—whether he took exactly this particular cross section or another one. Today, one can do all of this very easily in a simulation, but it is always the best strategy to continue thinking a bit longer before starting to simulate. There is no doubt that one will eventually be able to find, with the aid of a computer, that there indeed will be what Poincaré called a "homoclinic point." Then we have all of the complexity of chaos in this system in which we have perturbed the circular motion into an ellipse. At the time of Poincaré's, there were no computers, there was no chance to automatically plot this system which preserves all the complexity of the three-body problem. Today, because we know that we can take recourse to the computer, it would be impossible to do what he did by heart in his mind. His famous homoclinic point can be studied numerically to date. A schematic picture has once been called the "dollar attractor."

How could he proceed with this problem without a computer?

This is an important question for an historian of science. I would very much like to be informed about how he did it. Nevertheless I shy away from entering into the chapters and books that he wrote because it will be almost a life's task to extract the essence; how he realized exactly the existence of homoclinic orbits which is so maximally miraculous a finding in its transfinite complexity. He said that "the mind is cringing away in horror" from what it sees here. He had previously realized in his studies of 2-variable dynamical systems that there is a "uniqueness of trajectories" theorem in charge. With two variables in a plane in deterministic motion, we can only have continuous paths that never cross except in a so-called "critical point" where all motions stop.

The limit cycle is the second typical organizing phenomenon in two dimensions which he also discovered. Now, we have a 2-dimensional cross section through a locally 3-dimensional flow. Here we can, in addition, find crossing points as mentioned. These points follow rules. We saw already that all points created in the original circularly moving case lie on one line. They each determine an equal area. The phenomenon of "area preservation" that is familiar from frictionless two-variable systems remains valid for these maps. However we have much more complexity, once the circle of the third body is compressed into an ellipse. Then the successive crossing points show remarkedly complex behavior. Poincaré convinced himself that once more, all successive points must lie on a line in the cross section. They still lie on a line, not on one line per se, but rather on two lines simultaneously that continue to cross each other again and again, forever. The structure of how these two lines intersect and do this infinitely often in the plane of the cross section is what caused him to speak of a "horror." He saw that there still exist genuine critical points, like a saddle point. There also exist one-dimensional manifolds—lines—in the cross section which emanate from or tend towards a saddle point. To me it is a miracle that he saw this. Regarding these crossings, we have a new problem: On the one hand, these lines have to remain lines, on the other hand, they can now intersect. Imagine: You have only lines and each line is well defined but nonetheless now suddenly you can have an intersection. What is the matter with those points in which two lines intersect, if each point has to stay on its own line forever, one line exponentially expanding, the other exponentially shrinking? The two lines are bound to meet again and again, even though one of the two lines is getting longer and longer each time in between the crossings while the other is getting shorter and shorter. Both originate from a periodic point in the system, a fixed point in the map, which is a saddle point. The one impinging line (one calls it a manifold) exponentially approaches and the other exponentially departs from the saddle point. Hence a single intersection of the two entails infinitely many intersections. In this way an infinite complexity of criss-crossing lines arises—an incredibly "disciplined tangle," as Alfred Klemm would say. This finding led to the famous "stable manifold theorem" as Steven Smale called it. These crossing points, which forever spawn new crossing points, Poincaré called "homoclinic points"—points leaning towards the same point, namely the saddle point in two directions of time. They are not alone, there also exist "heteroclinic points."

Sorry, there are two questions here. The first is how can we calculate this Poincaré map?

I am reminded here of the fairy tale about the ghost of death who upon meeting a person is wondering why this person hurries so much to reach a certain far-away place which is exactly the destination at which he is scheduled to meet him for good in the evening. Forgive me, but the belief of Poincaré in the exactitude of all this is the real miracle. It turns out that this exactitude is not just infinitely exact; it is a transfinitely exact phenomenon. Poincaré lived at the same time as Georg Cantor who discovered the transfinite—trans-infinite—accuracy of mathematics, but Poincaré saw here the first physical example where it can be seen in action.

What is the meaning of transfinite or transinfinite?

Transfinite is a word coined by Cantor. He claimed and gave proofs of its trans-infinitely infinite magnitude, but no one believed him at the time. Namely, that there is a second type of infinity which is not just infinite, but trans-infinite in the sense of representing the infiniteth power of the former. This transfinite exactitude which was discovered by Cantor in the nineteenth century had already been seen by Anaxagoras around 500 B.C. in ancient Greece. In the intervening centuries and millennia, almost no one had come back to it, except apparently for Giordano Bruno, as Elisabeth von Samsonow told me. Then virtually at the same time Cantor discovered his transinfinite infinity and Poincaré gave the first example in a deterministic continuous system, a three-body system. That was the birth of modern chaos theory, the re-birth of Anaxagoras' insight. Chaos theory is a transfinitely exact theory. The Poincaré map is the best example so far. Poincaré was overwhelmed by the complexity of the homoclinic tangle as mentioned, speaking of a "horror" that befell him. Unfortunately, he did not appreciate Cantor, nor did he appreciate Boltzmann even though both were fishing in the same mathematical waters, so to speak. Three exceptional people working on the same problem after a gap of 2½ millennia.

Boltzmann used the word "chaos" but it did not have much connection to the modern idea?

It actually has, I believe. This model which we just looked at—the three-body problem with the infinitely disciplined tangle—actually is related to Boltzmann's discovery of the "hypothesis of molecular chaos" in statistical mechanics. It lies at the root of entropy formation, as far as I can presently see, but this is an area under construction right now.

This principle of "chaos" in Boltzmann's mind, why did he call it "chaos" in his phrase "hypothesis of molecular chaos"?

He saw in the statistics of many particles that interact deterministically that you can take an initial condition which is freshly picked so that the respective particles cannot have interacted as of yet. This is what he meant by "chaos"—that a particle is absolutely out of correlation with all others. He needed something like this because, if they had already interacted in the past, they would—if you invert time—go exactly back to that point where they started to interact. If you went further in that same time direction, from that moment on they would do in the backward direction of time what they did in the forward direction of time. The in between "virgin point" where nothing had happened so far in either direction of time cannot have had any history itself. It comes from a primordial chaos, as it were. It is quite possible in my mind that this primordial-chaos like state is related to Poincaré's deepest insight. At least it is related to Anaxagoras deepest insight that started out with the "perfect mixture" which he called "chaos." It is a great pity that Poincaré and Boltzmann never met. Poincaré was tragically repelled, saying about Boltzmann, "I do not read the papers of a colleague who starts out with an axiom and arrives at a theorem that contradicts his axiom." Boltzmann indeed started out with the assumption of reversibility (in the motion of atoms) and came to the conclusion of a time's arrow (irreversibility) on the macro level. He emphasized that this "irreversibility" holds true in both directions of time, this Poincaré could not know because he had not read Boltzmann.

Somehow the Poincaré recurrence was an idea designed to kill Boltzmann's deterministic time's arrow?

Not quite, but it of course actually is compatible. They really were friends in their minds, but they did never meet to find out.

This Poincaré section that you mentioned has a special power to it. Many people use the Euler method in the numerical analysis of a continuous system. If you discretize a continuous system which is chaotic using this linear approximation, then the discretized version often shows no chaos. The only reliable technique is to calculate a Poincaré section through the continuous system?

Yes, the Poincaré method of looking at a section is very general and does not just apply to the three-body problem that we looked at, but can be applied to any deterministic dynamical system of three variables. Even then it must be done accurately, as you say.

If you agree, I suggest to come back to the two-body problem. In the two-body problem, on his way towards the discovery of general relativity, Einstein hit on a phenomenon called rotation of the Mercury perihelion. The Mercury perihelion somehow shows quasiperiodic behavior rather than the periodic behavior of a Kepler ellipse?

This is correct. It is an ellipse which is a little bit open and hence is slowly rotating. So the motion forms a rosette eventually, and this never stops—it is a typical quasiperiodic motion.

In two-dimensional dynamical systems we have a special theorem called the Poincaré-Bendixson theorem, according to which we can only have periodic behavior but no crossing trajectories with two variables?

Yes, the three-dimensional behavior includes quasiperiodicity and chaos, and the relativistic two-body problem shows quasiperiodicity.

In the Mercury perihelion problem we have a two-body system showing quasi-periodic behavior. What is the new dimension in this system?

This is an interesting question, but the answer is a little bit trivial. If we write down a Hamiltonian of the Newtonian type for two bodies, in which the Newtonian attraction is perturbed a bit so that the force term is no longer quadratic in the denominator, but is perturbed by an added constant term, for example, then we no longer get the closed ellipses of Kepler but a rosette phenomenon. So virtually any nonlinear perturbation of Newtonian mechanics suffices to get a rosette. General relativity is not a particularly special case in this respect.

Post-Newtonian behavior?

You can say so, but this is unspecific. "Non-Newtonian behavior" would be a more general description.

The attraction function is just a little bit changed?

Yes, we do have four variables. The Kepler ellipse turns out to be structurally unstable in a sense. This particular structural instability deserves an investigation in its own right. It is not a phenomenon on the way towards chaos, although even this may still be an option yet to be ruled out.

Can we have chaos in a two-body system in which one is a black hole and one a point mass?

So far we had stuck to Newton and there we cannot have chaos in a two-body problem. But as soon as we enter general relativity, everything is much more complicated. I would guess that this chaos question has yet to be answered. I remember quite complicated unbounded motions in a picture displayed in the Misner-Thorne-Wheeler "bible" titled "Gravitation" of 1973. So my intuitive guess regarding your question would be "yes until falsified."

Some researchers tried to respond to this question after adding some other postulates or other new things, for example gravitational waves, and using these they tried to find a positive answer. We need something that does not involve additional variables. The gravitational waves do form something like an open question?

Yes, Einstein himself was doubtful regarding the existence of gravitational waves.

Let me conclude this subject with the following question; what do you think is an open problem in the structure of three-body systems that one can focus on to better understand it and produce something new? Or is this a closed research area?

It is an absolutely open research area even though it is so old. We here have to return to this big job of Poincaré's. One will have to re-study how he arrived at his beautiful transfinitely exact tangle of homoclinic points (and heteroclinic points) because this tangle apparently underlies the time's arrow of statistical mechanics, in either direction of time. Statistical mechanics as is well-known explains our macroscopic phenomena like heat production or evolution. It can now be replaced by this 3-body problem of Poincaré's on the most basic level. If we go from the Newtonian attraction considered so far to Newtonian repulsion (including more short-range repulsive forces) then we get statistical thermodynamics. If we stick to the attractive case, on the other hand, we unexpectedly get a new science which is a close analog to statistical thermodynamics. It too is a sub-discipline of statistical mechanics. In this new discipline, we do not encounter an entropy increase when the particles interact, but rather an entropy decrease. Thermodynamics then stands no longer alone, there is this sister discipline called "cryodynamics" which actually directly follows from this particular example of Poincaré that we tried to look at. So now there exists a direct continuation of Poincaré's work that leads into a second major subdiscipline of statistical mechanics besides thermodynamics called cryodynamics. This new science really is of a high rank. It happens to explain as a side result the phenomenon of the gravitational redshift in cosmology—since here, too, we have a purely attractive interaction between the involved particles, namely moving galaxies and photons. In light of this, Poincaré is absolutely up to date. We indeed need to continue right where he left, in order to understand modern phenomena like the Hubble redshift law. What is presently open is only the exact quantitative relation. It might still happen that it is only 50 % of the cosmological redshift described by the Hubble law that is explained by cryodynamics. However, it is much more likely—a priori speaking—that the beautiful transfinitely exact theory of Poincaré explains everything perfectly so that we do not need the expansion postulate at all. For it is very rare in science that the same phenomenon is caused by two radically different theories running in parallel—especially if one of the two involves not a single ad hoc postulate while the other needs 24 (as I recently counted—but I admit I am partisan here). Having encountered a qualitative explanation of the sophisticated Poincaréan type is just too promising an alternative. Maximally simple physically and nonetheless maximally sophisticated mathematically—almost a dream.

Let us finish this session with the following question; why did Einstein almost never talk about what we today call chaos or three-body systems? There is one document which is titled "Two-body problem in general theory of relativity," but this is not a full paper, only a letter that Einstein wrote as a response to a colleague who had tried to show that general relativity does not hold true in a special case. Einstein sent that letter to Physical Review to show that the colleague had not been right.

Einstein had been very much influenced by Poincaré when he was young, especially by the book "The value of Science" (La valeur de la science) which appeared in 1904 (with Flammarion). This was a year before Einstein finished his special-relativity paper. Einstein was very much impressed by Poincaré, but he knew, I believe, that he had to keep his own work steered clear from Poincaré's work, because Poincaré was so overwhelming—and had known virtually everything that Einstein had seen in special relativity except for two specific points in Einstein's paper, the transversal Doppler effect and ....

The twin paradox.

Exactly these two. But Einstein also had a maximally ingenious and enlightened pictorial insight, namely the single light ball expanding about two mutually receding observers. To me, this is his biggest overstepping of previous thinking. Both Poincaré and Einstein are miracles in themselves.

Maybe there was a natural repulsion between these two minds?

No, I would rather say they were "too close." So it was too dangerous to mix too heavily with what the other had done. Then the attractive force would have been too strong. Their minds were very close.

Is there a further point you want to add?

Thank you, but I am very happy that I barely survived your difficult questions.

Chapter 15
The Brain Equation

Let us come back to the brain equation. What is the role of nonlinear dynamical systems for understanding the brain, and can we have a brain theory in this framework?

This is a major scientific question. Is it possible to use the arsenal of physical, mathematical and technological thinking to contribute to biological progress? Biology by definition does not have a very close contact to simple mathematical thinking, although Darwin was a mathematician in his soul, as Konrad Lorenz and I found out in discussing the fundamental questions in biology. In biology itself, the fact that eventually one has to arrive at a mathematical understanding is not well known. Only my friend Robert Rosen represented the opposite stance; he had studied mathematics in order to understand biology, to him this was the royal road in biology. It is true that a notion like the "brain equation" is repugnant to biologists because they think the brain is such a big thing and needs so much devotion on so many levels of research, and also there are so many different types of brains, that it is impossible even to think of a single line of reasoning existing that could explain the brain in an equation. If that is meant to be a real suggestion, one has to be very apologetic and modest in presenting the idea.

So you believe that you can make a theory of the brain based on nonlinear dynamical systems?

Yes.

In this theory you only need classical physics? Or do you need quantum mechanics as well? Primarily, one has to start out from what one sees, that is, one uses classical thinking, although eventually classical thinking breaks down as it did in the history of physics. The question that Darwin first tried to understand—the mechanism of evolution (and the brain is a product of evolution)—is a purely classical topic. Is the brain equation intended to make a new theory of the brain?

The very term "brain equation" is almost a slap in the face of everybody who has not heard the word before. Unfortunately, this is also the reason why virtually nobody knows about it because it is almost too grandiose a term, even though this

A. Sanayei and O. E. Rössler, *Chaotic Harmony*,
Emergence, Complexity and Computation 11, DOI: 10.1007/978-3-319-06781-0_15,

2-word notion is virtually the only way to capture the content of the idea. Moreover, it is not a bottom-up approach, as most other biological progress is; it is a top-down approach. It is the same thinking which Darwin introduced into biology, although for some reason the latter's theory is not seen as a purely deductive paradigm in biology. The Darwinistic way of thinking seems still foreign to the profession. In his old age, Darwin once complained that he could no longer enjoy both formal mathematics and beautiful music—a fact which shows that he was acutely aware of the existence of these two realities. If you wish, he subcutaneously injected his mathematical explanation of evolution into the whole biological community so that they all got infected without realizing the nature of the disease. The argument is maximally formal. If one assumes that the environment is changing in an unpredictable fashion, Darwin showed by way of proof, then one automatically needs a source of endogenous change inside the organisms living in that environment, since otherwise, the species is bound to die out. This is a purely mathematical type of argument.

In a previous session, we talked about the structure of the brain equation. Here I try to shift the topic towards some consequences of this concept. The equation involves some variables. Do you need a special space-time definition for this equation, or do you just consider the usual classical space-time?

It is "naïve" in terms of physics, this optimality equation implicit in Darwinism. It is once more nothing but a minimalistic attempt to understand survival. "Survival" as introduced by Darwin is a very down-to-earth notion. He looked at changes in the environment and at what had to be the unavoidable minimalistic response demanded of a species—a biological surviving structure—if the environment is changing and the survival is to go on. He realized that there must exist an "endogenous change" which in the last instance compensates for the external change as far as its effect on survival is concerned. So he could predict that there is a random change occurring in the genetic (as it came to be called later) structure of the species. Unfortunately, he hereby had to include all the past histories—the fact that the past of different species is very different. So his principle of inheritable sudden changes, which we now call "mutations," does not include the claim that they are predictable in their individual features. For they depend, in order to be effective, on the whole pre-history of the organism in question. The survival of a species under a condition of temporal change of the environment thus automatically implies that all past changes that led to the biological structure in question enter into the potential survival-relevant effects of a new mutation or random change of design. This makes survival in the face of a slow temporal change of the environment a different (much more difficult) problem to solve compared to survival with respect to a shortest-term change in the environment. The latter class includes spatial changes, in the specific sense that something that is momentarily encroaching upon the organism has to be fled from, or, if the close-by resource is a positive one like food, has to be approached. In other words, survival here depends on the organism's momentary position in space. This is an absolutely Darwinian-in-spirit question and can be answered just as he did with his slow "climatic" changes as it were. It "only" turns out as a big surprise that in this case, the prehistory—the chemical and biochemical history—does not enter in an essential fashion. If an organism's survival is impeded by something that is coming close to it, then it is a necessary strategy that the organism leaves the present position in space, or vice versa in the case of a close-by bonanza. This prediction just made is independent of the chemical prehistory of the species. This situation actually turns out to be a well posed optimality problem in the sense of mathematics. Eric Charnov called this way of thinking "optimal foraging theory" since this is what is required here, not to make lethal mistakes in one's changing position in space. If you forage, you survive in space—for example in a forest (forgive me the pun). In other words; when survival depends on position in space, then all you need do is make sure that a change in position improves survival. This is a mathematically well-posed problem. It can be solved only by mathematical means, by not violating the mathematically given opportunity. The result of this insight is the brain equation, so to speak. The brain equation applies beyond a certain level of urgency as it were (simpler solutions suffice if the environment is replete with appropriate sources to harvest, for example). So it is a very trivial approach to the problem of survival, much as Darwin's own approach was in spite of its incredible power. Both approaches are necessarily and trivially correct. Nature is bound to obey the mathematical insights which one finds in this way.

The brain equation is "top-down" which at first sight means it is unrealistic; nevertheless it is absolutely cogent in the sense that it cannot be avoided if you try to understand the brain. You suddenly can see why you have to look for what we call a "brain" in biology, and when it is unavoidable to have a brain. You can predict the properties of a machine that is needed for survival, for position-dependent survival. You can call it a brain even if you never saw a brain before, because you can predict from first principles what has to be done by nature if survival depends on position in space.

Can this equation also explain diseases of the brain?

Do you mean metabolic diseases of the brain, or infections? Or do you think of functional ones like schizophrenia?

Can these diseases be described in terms of variables of the brain equation?

Once you have implemented the brain equation with a special hardware, like a vertebrate or mollusk brain, the possible diseases depend on the particular realization. The brain equation is too general to help much when you come to particular diseases, although certain function-change type diseases become predictable.

Another question; in this equation, do you adopt a holistic viewpoint? Do you consider the brain as a whole and try to explain it and do not try to specify its different parts?

It depends. This is a purely functional approach; it has nothing to do with how this function is implemented hardware-wise, although you can also make predictions about its implementation eventually. First you have to know what the brain is doing, what is the essential function without fulfilling which it would not be a brain? When you have the blueprint of the function that this machine has to fulfill then you have a decisive entry into understanding the brain. It is just a footnote to Darwin for short-term spatial and temporal dependences of survival.

In 1994, you wrote a paper entitled "Microconstructivism." Here I want to know what is constructivism? Especially, since you mention in that paper the relation between "holism" and "constructivism." The paper addresses the brain equation as well.

That paper has to do with my more general idea of endophysics, which means that you look at the whole world from the outside. It is again an unfamiliar type of thinking, much like the spatial Darwinism we just talked about. Copernicus had adopted a similar way of looking at the solar system from the outside in the footsteps of Aristarchus. So it is indeed possible to have scientific insights which do not follow the usual path but give you a different angle of attack as it were. Endophysics means that you can look at the whole physical world, at everything in the cosmos, from the outside and thereby better understand what applies when you are inside. When you then watch the poor creatures creeping along inside this prison, you realize that from the inside of that prison—like in Marxism—certain things cannot possibly be understood. When you look at the situation from the outside, you understand what cannot be seen from the inside. "Microconstructivism" would be the right word referring to the new micro phenomena which predictably appear a-causal if you are a part of that world, but can be explained and predicted causally if you look at the situation from the outside. Thus, it is an explanation of quantum mechanics which is attempted by microconstructivism. If you wish, this is the next, more fine-grained way of looking at the whole thing—the next level after the brain equation. The brain equation itself is more restricted because it deals only with the survival of an organism, which is a simpler problem than the measurement problem of physics.

In that paper you say that constructivism describes a pseudo-reality. Do you consider the brain equation as an endophysics-type equation? If this equation is not valid from the point of view of an observer watching the brain from the outside, then your brain equation describes a pseudo reality; correct?

Yes, we cannot see the real truth from the inside, but the brain equation is much more trivial. We can safely return to ordinary—single-level—thinking. It is ordinary simple Darwinian science where you look at a phenomenon like the survival of an animal—comparable to Buridan's proverbial donkey which, placed between two sources of food each at the same distance, was claimed to starve because he could not decide which one to choose since both were equally attractive. The brain equation is exactly this type of thinking. Buridan invented it. It is naïve, mathematically cogent, and Buridan's donkey would actually be saved by the brain equation as a slightly more sophisticated step in the same type of reasoning.

Can the brain equation be improved from the endophysical viewpoint?

I do not think so. This is a macroscopic, classical, dissipative science—biological science in the traditional Darwinian sense. Endophysics tries to explain the scandal of quantum mechanics itself which abruptly ceases to be classical as is well known. So it is a different type of question that is addressed in the two approaches.

But in the mentioned paper you say that microconstructivism is a theory of the brain—and now you have a brain equation. So what is the relationship between micro-constructivism and the brain equation?

Thank you for hitting the point. The endo approach which describes the whole world from the outside, finds out surprisingly that certain microscopic properties of the brain affect the way the world looks if viewed from the inside. We do look at the world from the inside, and we do have a brain, and some features of our brain can be shown to change the way the outside world appears to us. What are hereby affected are only micro motions. Let me give a simplified example. If there were a very fast microscopic oscillation going on in the brain when looked at from the outside (exo) perspective, then we would not be able to "edit-out" this oscillation from the inside. Therefore, the world—the rest of the world—would be oscillating for us even though in reality, only our brain would be oscillating. This is because only the difference between those micro features of our brain and the rest of the world would be accessible to us. This is the endo perspective. Analogously, if there were a classical chaotic noise present on the micro level in the brain—objectively or "exo" speaking—we could not edit this noise out. This finding can in principle explain quantum mechanics, I believe. However, all of this has only to do with the microscopic physical implementation of the brain. Endophysics here is an ultimately inescapable way of looking at the world from the outside. Being microscopic, endophysics is independent in theory from the brain equation as the latter is implemented in a particular macroscopic brain. Thus we encounter the usual distinction between macroscopic dissipative physics (brain equation) and the microscopic, time-reversible and by a factor of 23 orders of magnitude more fine-grained, micro reality addressed by endophysics. While it is true that some of those lowest-level features become manifest on the macro level—as quantum mechanics is living proof of—the macro machine called "brain" can itself be consistently described without taking recourse to the micro level.

Can this equation show chaotic behavior?

Yes, it does. I did not know this when I presented the brain equation in 1973. The proceedings appeared a year later. I had felt obliged to add a mechanism that I named a "will-o'-the-wisp potential" to generate the requisite randomness in the behavior produced by the brain equation. I did not know at the time that the brain equation in its original form does already have this chaotic feature built-in, due to its nonlinear dissipative dynamical character.

What is this element of chaotic behavior? What is the meaning of chaotic behavior in the brain equation, actually?

The brain equation is essentially a "motivation system." A colleague of Wolfgang Köhler, Kurt Lewin, in the 1930s invented a "topological psychology" which is related (with its "vector forces") as I later learned. The brain equation describes attraction and repulsion as a function of the organism's position in the environment, and as a function of time. Specifically, it contains a function of the time elapsed since the last encounter with a filling station of a given type (since the individual types of "tanks" or alarm-clocks need to be refilled or "reset" at specific locations scattered in space). The "travelling salesman with alarm-clocks problem" is quite intuitive. An artificial device or an animal employing the brain equation therefore automatically follows a chaotic course in the environment. This chaotic behavior is sensitively dependent on initial conditions. So it is indeed possible to say that the brain equation implies chaotic behavior.

A brain that works in the normal sense should accordingly behave in a chaotic manner?

Correct.

Did you ever think that some of the diseases that were mentioned—like MS—might lead the system from chaotic towards non-chaotic behavior—towards periodic behavior, say? And would this insight open up the possibility to change the brain back into the chaotic mode?

Eventually, such ideas have to be considered. Multiple Sclerosis probably would not be a very good example though, because it is an anatomical disease where certain structures in the white matter of the brain are essentially mechanically interfered with. Nerve fibers are interrupted in their function by inflammatory processes so as to no longer work properly. Such a dysfunction depends on the particular way in which this brain obeying the brain equation has been wired. The same disease may not exist in an arthropode like a mantis shrimp which likewise implements the brain equation and is a very intelligent animal. Or think of a mollusk like the giant octopus, which also has reached a high level of implementation of the brain equation (including an added universal simulator that makes it arbitrarily intelligent in principle).

As I learned, biologists do not easily take to the idea that physicists, or people working in dynamical systems, claim they can apply a perturbation to the brain and thereby change its function from a chaotic toward a non-chaotic mode or vice versa in order to cure a functional disease. Why do they not wish to have a part in such a proposal?

Physiologist Horst Prehn from the University of Giessen once did what he called the "braindrops" experiment. He let a human person watch a chaotically dripping faucet, recorded the produced brain waves live with an EEG machine and fed these waves back via a transducer to the dripping faucet in question, as a mechanical perturbation in real time. The chaos immediately stopped—the faucet dripped only periodically from that moment on. This beautiful experiment is much too little known.

Why do we have this problem of a lack of communication between the two cultures as it were? From the mathematical point of view, it ought to be possible to bridge the gap. Why do biologists or physicians tend to not take physical or dynamical arguments seriously?

It is, I believe, a psychological problem—a strong hesitation to accept mathematics. Biologists and medical people apparently do not realize that in science, eventually everything reduces to mathematics. It is due to lack of information, I believe.

What are the open or future problems regarding scientists working with this equation? Why did you not work on this equation for a long time?

I talked about it in my courses and it still happens that students are very much interested. In the United States, Michael Conrad took it up; in Japan, Kunihiko Kaneko advised a doctoral student to work on the brain equation but as far as I know it did not get finished. I was invited to give a talk at Tokyo University two decades ago. It was a physics group and I lost contact for not being available enough as mentioned. I am very sorry for that. There are two cultures, physics on the one hand and biology on the other, each of them having to bridge a gap that the world does not know exists. The brain equation is a typical "no man's land" concept. It actually is pure mathematics, of a type totally different from other branches that are applied in the sciences. Garey and Johnson's famous 1979 book "Computers and Intractablility," written on the NP-completeness problem is the best intermediary approach, but it gives no connections, neither to biology nor to physics. So a whole big field is still waiting to be built up before the intermediary discipline of Artificial Intelligence can be expected to take the theoretical lead again in the footsteps of Turing.

Would it not be nice to find a group of some mathematicians, physicists and biologists to work on this problem and find some results? Would you support such a group?

By all means. Actually there is presently the biggest program in science ever brought on its way, funded with 1 billion euros by the European Community, with the aim of understanding the brain better, up to building an artificial analog. Its name is "Modelling4All." Of course, this program has never heard of the brain equation. The brain equation could no doubt help this group of interdisciplinary enthusiasts to find a common ground and succeed. There also exists an interdisciplinary book written with Bill Seaman as the first author, titled "Neosentience." A concerted way of thinking about the brain, with artists working hand in hand with engineers and scientists, is about to emerge at the time being. I should also mention the beautiful artificial face, first designed with 5 naturally combinable "emotions" by my pupil Wilfried Musterle 26 years ago. There is a recent heart-moving documentary of a mechanically implemented artificial child-face on the Internet. You see, the "top down" approach is almost on its way by now.

Do you think that this type of thinking manifest in the brain equation can be implemented in an artificial intelligence—to make some viable robots?

I even believe this is the only way to make intelligent robots in a functional sense, so that they are not just intelligent for certain tasks but are autonomous actors (in both senses of the word) that eventually become potential competitors to animal or human intelligences—unless we make peace with them and find a way of cooperation much as the different religions have to cooperate with each other. This task is also unsolved—despite the efforts of people like Mevlana Rumi in the distant past and Emmanuel Lévinas in the recent past. Miraculously, really lovable persons emerge from time to time. Now they could get support by a deeper scientific understanding of two brain equations in interaction.

Do you want to add any points because my questions are finished?

There is one point which I would like to add—the qualitatively different use made of their brains by human beings and other animals. It has to do with the interaction problem on which we just finished. We better let it float in the air here. Thank you very much for your questions.

Chapter 16
Repulsion and Attraction

In our new session we are going to talk about cryodynamics. Let me start with statistical mechanics. We have some postulates or axioms in thermodynamics, calling them "first law," "second law," "third law" and even "zeroth law." Altogether we have four laws that we can call axioms, but some of them can apparently be derived from one of the others. Some books start out with the axioms and try to offer the whole theory based on these axioms. One question which naturally arises is; why can we not change these axioms in order to come up with a complementary theory? For example, one axiom says that the rate of entropy change should be positive, and another axiom says no, it can also be negative. Can a complementary theory complete this? Many books on thermodynamics after stating some axioms go on with microscopically specified examples in order to then generalize them to quantum mechanics. Others start out with quantum mechanics, building the whole theory up from there to then find the limits under which we can go over to classical physics. Both approaches seem to be legitimate. Now let us start taking this as a backdrop, what is cryodynamics? Can we again start out from quantum mechanics to go to classical physics or should we depart from classical physics?

This is a well-taken difficult question. Everybody has of course his own private way of thinking with his or her acquired preferences. To my taste—and from my experience with chaotic systems that have but a few variables interacting—it appears legitimate to start out from a microscopic causal mechanism in a classical mechanical setting. Thermodynamics is the "Theory of Heat" with all its different aspects including the notion of entropy, and so on. It has its quantum analogs and special cases and explanations, but the door towards really understanding what happens is in my view, classical chaos theory. Looking at two or more particles—a gas has very many particles—is the starting point. What are the general properties when you have but a few interacting particles? In chaos theory, Yakov Sinai had the first model of chaotic billiard balls interacting in a plane. I was told by Harry Thomas in Basel in 1976 that he had found a way to visualize the chaotic interaction of particles, according to Sinai. As mentioned, he called it the "tennis game in an orchard"; a tennis ball flying horizontally through an orchard of regularly

A. Sanayei and O. E. Rössler, *Chaotic Harmony*,
Emergence, Complexity and Computation 11, DOI: 10.1007/978-3-319-06781-0_16,

planted trees. If you have a tennis ball flying—or rolling without friction—through the idealized orchard, then whenever the frictionless ball is headed exactly parallel to the infinite line of trees (the latter is assumed to be infinite), nothing will happen to it, it will just continue forever. However, as soon as you give this tennis ball the slightest deviation from the absolutely parallel direction in its alley, it is bound to eventually hit a tree and hence be reflected into a new direction with in general a different angular deviation from the middle line. From then on, it will no longer move along a privileged straight line but move chaotically. The other straight avenues that run through the regular orchard have the same property of being unstable with respect to the slightest deviation from straightness inside. So Sinai could prove in a long paper in 1970 that this system is chaotic—meaning that almost all initial conditions lead to total unpredictability of the path of the particle after a short time, even though everything is deterministic and can be understood in an absolutely classical fashion. The word "chaos" was not yet coined for this kind of deterministic random behavior (this would happen only 5 years later in T.Y. Li and Jim Yorke's paper "Period Three implies Chaos"). There are quantum versions of the Sinai phenomenon. Also, the particles, tennis balls, are ideally hard spheres, but you can smooth them at the boundary a little bit, so the balls become softer. Then still most of the results persist. Sinai's theory is the simplest example of thermodynamics, showing how chaos enters into the description of a deterministic system, with thermodynamics forming an implication. All the insights that chaos theory has brought are applicable to this picture, and eventually you can hope to derive all the other laws of thermodynamics from this paradigm of a "Sinai gas."

It is derived from billiards?

Yes. When you look at a two-dimensional section of the rolling billiard ball, you get a Sinai disk. Sinai first saw that you can replace the balls by two-dimensional disks. He then nailed one disk onto the middle of a quadratic table. Opposite sides of the quadratic table he identified in a pairwise manner. So the table became a torus, that is, all motions reaching the boundary would continue straight in the next "cell" which would be identical to the old cell. On this torus-shaped table, there was this round disk-shaped repulsive obstacle nailed to the center of it. The moving second ball he shrunk to a point, after doubling the central disk in diameter, without changing the former's path. The angles of collision with the circular obstacle in the middle would always be locally symmetric, that is equal from both sides. This model is the simplest example to explain thermodynamics; a billiard table with pairs of opposing sides identified, which is mathematically trivial to assume. It is the most impressive explanation of gas theory, whereby the word "gas" itself means "chaos." It was coined by van Helmont in the early seventeenth century—chaos and gas are the same thing. Sinai's gas is a deterministic explanation of the statistical behavior which occurs with many particles that are mutually repulsive. You can use this chaos-borne example to eventually understand, I would say, all of the essential oddities of thermodynamics, and you can also have a quantum version. Then it depends on your taste whether you prefer quantum mechanics to be the basic theory or whether you think that the world is basically classical. Quantum mechanics can in principle be explained from a classical ("endo") perspective in my view. Sinai's would then be a nice example of the theory of heat, thermodynamics, as a whole.

So you prefer to start from classical theory to explain thermodynamics and cryodynamics?

From classical chaos.

So what is the meaning of cryodynamics in your eyes? As a matter of fact, this word was created by you?

There are some manufacturers for low-temperature products carrying the same name. The name is therefore rather widespread, but in science, no one else introduced it so far.

Who was the first person to talk about the new behavior of cryodynamics? Give us at first a history of the subject and then, please, describe what cryodynamics is.

I should first perhaps make a little modification to the billiard type model of Sinai that we just looked at. You can smooth the potentials so that the disk's walls become softly repulsive. Then you have a variant of the billiard model that is more realistic. The chaos which they jointly produce according to Sinai then contains what is called "KAM-tori." They are embedded in a chaotic regime as little islands of many sizes that are not chaotic inside. They are quasiperiodic, with non-diverging trajectories internally. Nevertheless the quasiperiodic subset of motions embedded in the chaos can be neglected, quantitatively speaking.

You mean Kolmogorov, Arnold, and Moser?

Exactly: KAM. Kolmogorov was Sinai's teacher. I once had a telephone conversation with Jürgen Moser before publishing my paper "Different kinds of chaos" in 1976. He was a maximally gentle great mind. Here what we saw is that thermodynamics is still basically valid if the boundaries of the involved particles are smoothly repulsive, despite the fact that, in principle, there are now also these islands within which the system's motion is not chaotic (only quasiperiodic). The measure—more properly speaking the relative size—of these domains is negligible for all practical purposes in the multi-particle case. So this modified smoothed-out Sinai scenario still explains all of equilibrium thermodynamics, and we can still apply quantum mechanics as a special case which constrains some of the results. Thermodynamics remains explicable with its overall features in terms of these repulsive particles even if the repulsion between the particles is not of the hard-spheres type. Molecules in the air, for example, do repel each other. Their repulsive boundaries are very limited in their range, but you can ask the question; what happens if you make the repulsion more long range? In particular, what happens if you use a Newton-like law which is quadratic in terms of the force, yet repulsive rather than attractive (obtainable by a mere sign flip)? Then you still retain all the essential properties of thermodynamics if the particles are fast-moving, which fact is perhaps not very surprising. However, you can now also replace the repulsion by attraction, genuine Newtonian $1/r$ attraction (or an equivalent under more or less static electric or magnetic attraction). Although this may appear a trivial generalization, this attractive multi-particle case was neglected ever since Newton, which is very surprising. Nevertheless this case still makes sense. You can build-up a theory of many particles that interact not in a repulsive fashion as in thermodynamics but in an attractive fashion. This "sister program" is a very simple idea, and although some important people, like Lev Landau and Donald Lynden-Bell and my friend Rolf Landauer, wrote about this in the 1930s and 1960s and 1970s, respectively, only a combination of attractive and repulsive regimes were looked at in the past. This very artificial and complicated situation did not yield major new insights at the time as far as I am aware. The "pure case" which I mentioned—many attractive bodies in interaction—was apparently never studied in its own right. The reason why I am dwelling on this is that, you now no longer get thermodynamics; rather, you get a totally different type of qualitative behavior, that of cryodynamics.

One question, you said that the potential is attractive here?

That is the distinguishing difference between thermodynamics and cryodynamics.

Why do you say that it is a new idea? We have attractive potentials in Newton's old theory. So what is new here?

It is not at all new, you are right, Isaac Newton could have done it. This indeed is the oldest class of systems we have in exact physics. Nevertheless for some reason, the interaction between many mutually attractive particles like galaxies, for example, has never been looked at from a many-particles point of view, with an eye to the qualitative implications of this type of interaction. Chaos-type qualitative thinking would have been required. This beautiful new situation can be simplified into a Sinai-billiard analogous case again, only with an attractive rather than repulsive potential involved. Sinai brought the breakthrough in the repulsive theory, showing thermodynamics is not indeterministic but chaos-based. Since the chaos-theoretic breakthrough, again four decades had to pass before this natural attractive Newtonian analog came into focus.

So you believe that Rolf Landauer was the first to start to talk somehow about this?

Not really, Rolf was essentially working on repulsive thermodynamics.

Who was the next one? What is the history of this attraction-based viewpoint?

One has to go back much further, towards a different strand of science: celestial mechanics. The first here was, as far as I can see, Fritz Zwicky, a Swiss-American astronomer who is very famous because he discovered dark matter—baryonic dark matter, to be precise. He has many other ingenious ideas to his credit, like what humankind can do when the sun grows dangerously large in some 5 billion years' time from now. Then one could give earth a gentle repetitive push to move away from the sun for a sufficient distance, he showed. This will no doubt become more feasible as time goes by. In our context, Zwicky had the association in his mind that when a photon is travelling through the cosmos, the cauldron of the randomly moving galaxies it is passing through would influence the photon's energy in a systematic fashion. That was a qualitatively new idea that no one else had had before. It is the first example of cryodynamics in action—to look at the interaction of many particles that attract each other, galaxies and photons. The photon is attracted, of course, like any other particle. If you wish, you can replace it with a cosmic-ray particle, a proton of almost the same speed as a photon. The fast particles in question then interact gravitationally with the many moving gravitation centers, called galaxies and stars and so on, encountered in a grazing fashion. Thus, we have the whole sky as an implementation of the idea of cryodynamics. Zwicky was the first to see the principle. So I have a very high respect for Zwicky—that he was able to visualize this as early on as he did. Unfortunately, he did not follow up on this insight with full vigor. He got discouraged by another great scientist, Sir Arthur Eddington, who has many creative contributions to his credit, too. Eddington found an error in the equation that Zwicky had published and Zwicky had the courage to quote from Eddington's private letter to him in the next publication and admit that he had committed a mathematical error. So the whole world was laughing and Zwicky had to make a new big discovery to regain his credit as it were. People who are very creative—like Zwicky or Maxwell—are notorious for making mistakes because if you are not afraid of finding something new, it is virtually impossible not to make a mistake nine times out of ten. Only by taking risks and then finding out together what was wrong can one make progress in science. Ideas must not be killed before they have gotten a chance to be proved, not in their details but in their essential core. This probing process can take more than 7 decades as in the case of Zwicky.

After this drawback, he never talked about it?

He wrote a few more papers on the subject which I admit I did not check in detail. He never gave up but he somehow was discouraged enough not to invest his full force to help the idea up into a form in which it could find acceptance. Imagine: "Tired light!" as it was called later. This epithet could not be washed-off. Nevertheless the idea went through a second eye of a needle in history. His great adversary Eddington had a pupil named Chandrasekhar. He came at age 19 to England from India. He famously had a good idea on board the ship to England, namely the quantum idea which implied that white dwarfs can exist due to electron degeneracy, thereby paving the way also toward the later theory of neutron stars—begun by Zwicky—and even of black holes. Chandrasekhar then wrote a paper in 1943, 14 years after Zwicky on what he called "dynamical friction." It essentially reproduces Zwicky's result with a different method but does not mention Zwicky. Nevertheless he gave a name to the phenomenon that Zwicky had discovered less elegantly (and with an error) before, calling it dynamical friction.

Dynamical friction means that when you have a particle travelling through a cloud of attractive particles in random motion, this particle will be slowed—lose kinetic energy—as if by friction, rather than gain in kinetic energy statistically as everyone would expect in the wake of thermodynamics. The effect looks like friction even though there is no collision anywhere here—it is all Newtonian forces acting at a distance. This phenomenon Chandrasekhar discovered and proved painstakingly in his 1943 paper, but so without seeing its whole significance. He not only missed making the connection to Zwicky, he also did not realize that his braking phenomenon did not depend crucially on the mass of the braked particle as one is familiar with from statistical mechanics. That is, he still had a Maxwellian velocity distribution in his mind in which overly fast particles get reduced towards the equilibrium kinetic energy to explain the braking effect. He looked at a system of stars, a special type of star clusters called "globular clusters." These star clusters are very pretty, having a very pointed density maximum in the middle. He realized that in the maximally dense core, from time to time an "almost triple collision" occurs. Stars are too small to genuinely collide often, but quite frequently three stars meet in such a way that two stars come momentarily close in a very pointed Kepler ellipse while simultaneously a third star happens to be passing by. Then the two of them can shed their potential energy giving it to the passing third star in the form of a very large amount of kinetic energy. The accelerated star then is shooting out from the cluster's center leaving behind two closely bound stars called a "star molecule." The fast star tries to leave the cluster shooting out from the middle. What Chandrasekhar recognized is that if there did not exist a braking force preventing the accelerated star from leaving the star cluster, these globular clusters would long have ceased to exist. However, they actually are the oldest known structures in the galaxy and even the universe (as mentioned). Chandrasekhar was the first to explain this amazing empirical stability. He could show that a star racing out from the center of this cloud of stars will be braked enough to remain within the cluster for good. But having confined his interest to stars, he did not pay much attention to what happens if the fast particle is not a star but a moon or an equally fast asteroid, etc. Nonetheless he conscientiously showed in an inconspicuous quantitative note that there is still almost the same braking effect going on for lower-mass particles. Nevertheless, he forgot about this in the rest of the paper, quoting Maxwell's distribution of ordinary thermodynamics, as mentioned. Obviously he never realized that here lay a big contradiction, and that he had described a completely new theory from the point of view of statistical mechanics. Since his method had been a Bownian-motion type stochastic analysis in the footsteps of Einstein and Smoluchowski, the contradiction did not force itself onto his mind. The textbooks in "Galactic Dynamics" continue reproducing this error—despite the great learnedness of Binney and Tremaine's book in all other respects. It was a pity that Chandrasekhar apparently never met Zwicky to talk shop with him.

You mean that in this case statistical thermodynamics cannot be applied? Why so?

It is because the laws of the repulsive smooth Sinai billiard and the attractive smooth Sinai billiard are totally different. The attraction-based theory sports an almost perfect mirror symmetry to statistical thermodynamics. Virtually everything still waits to be formulated out. This attractive version of statistical mechanics has essentially been overlooked since Zwicky. It embodies one of the fundamental questions that should have been asked since the beginning of statistical physics and many-body celestial mechanics. So it was possible to discover a big new branch in the foundations of physics. When you fill it out, a whole new science emerges that is just as big as venerable old statistical thermodynamics—namely statistical cryodynamics. It is beautiful to have a second major physical theory of the same formal richness as one of the main pillars in the fabric of science. It takes a little time to work out all the implications. In comparison, it took quite a few decades until the steam engine came and eventually got morphed into a small gasoline or diesel engine for everyone's use.

What is your contribution in this context?

Very minor so far. I am only one of those who recognized the existence of the new science and dared give it a name.

But you worked on it for several years?

The idea arose in a lecture course that I gave on chaos. I had students who attended this course repetitively for the fun of it. One of them is Dieter Fröhlich, another was the late Normann Kleiner who was my age teaching at a gymnasium. In our discussing in the lecture hall, one of us—each thinks it was the other—came up with the idea of asking what would happen to a light particle if it were to pass a large number of mirrors in random motion. This idea occurred in the context of looking at the sky with a cosmological bent. Is there a lawful distance-dependence of the speed or the energy of the particles that are traversing the universe's cauldron of moving galaxies? The Hubble redshift was the reason that we arrived at this picture of "mirrors" moving about. So we used in our minds a repulsive model, which is just the same as that of statistical mechanics, and asked ourselves the question of whether it was not also applicable as a model to the cosmos with its cosmological redshift. As already said, the sign of the involved forces was wrong since we assumed repulsion rather than attraction. So the original idea was radically false but nonetheless helped us to see the other question—what happens when you have many attractive heavy particles moving around like the previously assumed mirrors, and a light particle is moving through? Thus, even though the repulsive mirrors were misleading, the idea of the photon losing energy while traversing many moving particles in the cosmos proved fertile. So we re-discovered in this roundabout manner the very question that Zwicky had already formulated in 1929, 73 years later in Tübingen. I gave a talk on this at the University of Oldenburg the same year. Our chapter appeared a year later—without the error of sign—and we published quite a few papers over the years in which we elaborated on this idea—without realizing its more general significance at first. I mean, we saw that this insight might put into question the expansion model of the universe because presently the photons are still falsely believed—as Zwicky first tried to debate—to lose their energy along their journey through the cosmos through a global space expansion.

You focus on one photon in your model. If you have two photons, for example, or very many: what is going to happen then?

We have many galaxies and we can also have many photons, but a single photon suffices to understand the essence.

But if you have two photons then an interaction between them is possible, there is a kind of gravity between them?

Yes, but this effect is negligible here compared to the effects of the big masses around.

If we want to go to quantum mechanics since you said that we can apply this idea to quantum mechanics as well; would this gravitational effect in quantum mechanics between two photons become important?

Yes. This question will become significant when we have a very high photon density going through the cosmos, which no doubt is the case in some regions. John Wheeler even invented the notion of "geons"—a torus-shaped cloud of photons that are held together by gravitation. It is unfortunately unstable, but, if you add quantum mechanics…

…we should then talk about the interaction of photons and matter?

Presently, it is the low-energy photons that are being cooled by the high-energy galaxies. This process of "paradoxical cooling" is the characteristic of cryodynamics.

How do you enter temperature into your model?

Temperature is a notion from thermodynamics, it is the mean kinetic energy of particles. We can look at situations in which we have not photons and galaxies—which was our only example so far—but in which we have, for example, a gas of high-temperature nuclei and electrons. This is called a hot plasma. We can send through this hot plasma a ray of even hotter electrons. For example, we can assume that we have a very hot plasma of atomic nuclei and that we are embedding this plasma into an even hotter plasma of electrons, which one could generate artificially without many problems. All one needs is to use several electron accelerators (six in the Chu scheme) and concentrate their beams onto a hot spot within the plasma of nucleons. Then this local electron gas of a higher temperature will predictably cool the plasma of hot nucleons. Even though this is only a gedanken experiment, a thought experiment, it is applicable to real-life hot plasmas. It apparently represents a solution to the problem of stabilizing ITER—a machine which is being built for 3 decades and is expected to need 3 more before being ready to generate inexhaustible free energy through hot fusion by pulling the energy source of the sun down onto earth. This machine is not expected to work soon, but cryodynamics could potentially be used to make it work sooner. Cryodynamics would then have its first down-to-earth application. Zwicky's rebuttal of the Big Bang would then no longer stand alone, and his idea of tired light would become an applauded household name. The Big Bang is rather likely to become a by-gone pure mythology in the history of science much like the famous phlogiston—if cryodynamics is confirmed as a fundamental principle of physics. Cryodynamics would prove to be more than an abstract theory of the same beauty as thermodynamics by becoming the motor of world economics just as thermodynamics had done before.

What is going to happen to the entropy then?

You are touching on the abstract theory of thermodynamics in which the most important technical notion is entropy. We know, for example, as Erwin Schrödinger put it, that "life is feeding on negative entropy." So entropy is very important and life would not exist if entropy was not such a reigning phenomenon. Ilya Prigogine devoted his life's work to this. Also Teilhard de Chardin and the whole of evolutionary theory is based on thermodynamics. Now cryodynamics has surfaced as a complement to thermodynamics. All the important notions that we have in thermodynamics—like entropy and far-from-equilibrium "dissipative structures" including flames and life itself—are thriving on entropy generation. All of this then no longer stands alone. In cryodynamics, we have a radically new notion which is called "ectropy." It is numerically equal to entropy—except for sign. You can again write down equations here. In thermodynamics, Hans Diebner started doing that in Tübingen with a deterministic entropy formula. There might even be an analog to life within the scope of cryodynamics. If so, it would be much more vigorous because life is just running on the back of the global entropy increase in thermodynamics. By contrast, the "anti-life" would be an implication of the main force at work—ectropy increase—which generates the higher order directly.

So in ectropy the rate of entropy production is negative rather than…

Yes, ectropy is entropy with the opposite sign.

Now a question; people are trying to define time based on entropy generation…

…Like Boltzmann.

If we now have another role for time based on ectropy generation, would this allow time to go backwards toward its origin?

So far, we have no way to change the flow of time. Actually one can prove that time does not flow at all, but we have the impression of time flowing because we find ourselves in a new slice of time at every moment, a fact that cannot be explained by physics. No one knows where the "Now" is coming from, the Now that forces us to move through time along with the increasing entropy. Cryodynamics might eventually encourage someone to find a way to do something against this inexorable fate. This is a most elegant idea of yours. Even death could lose its sting.

My last question; if we go to the quantum version of this new physical theory, do you think that we then have a new phenomenon once more?

I would dare predict that the answer will be yes. At the present moment in time there are many easier to derive classical implications. But who knows; maybe if one starts out from quantum mechanics, one can perhaps arrive faster at novel implications of cryodynamics. The old intuition of Boltzmann, turned around, could be an avenue worth embarking on. Using quantum mechanics for finding new features of cryodynamics would be an alternative way to proceed. I remember Garnet Ord showing a slide of the double slit experiment with Feynman paths underlaid, with half of them drawn in red, the other half in green. Green was the involved time-inverted motions which crucially contribute to the two-slit phenomenon of quantum mechanics. So, on the fundamental level of quantum mechanics, we have two time directions acting in harmony together. Cryodynamics and thermodynamics might constitute a reflection, a twin pair on the macro level.

Also the photon is going to be very important in this context. Do you think that if the rate of entropy is decreasing rather than increasing, we can somehow conclude that heat can be transported from a lower temperature to a higher temperature rather than vice versa?

Yes, that is actually the essence of cryodynamics.

Do you have any point that you want to add here?

I would just like to thank you for the expansion of the field that we achieved in this discussion.

Chapter 17
Einstein

In this session we decided to talk about some special names in physics. We could start with Newton, but let us start out with Einstein. People love to know more about him. I am sure you studied a lot regarding his life, his paper and maybe his errors, if any. You can start from whatever point you wish. Maybe it is better if you mix both the scientific picture of Einstein and his life, and the papers and potential errors….

He did not commit any errors.

Actually there exists a book called "Einstein's Mistakes" by Hans Ohanian. Einstein himself believed that there are incorrect papers under his name. Everyone has this right to err because we are human. Please, start out from any point you like to talk about Einstein.

Every child on the planet knows about Einstein, and everybody somehow loves Einstein. There is the tragedy in Japan that people cannot accept Einstein as of yet as a person who deserves devotion because of the fact that he is indirectly co-responsible for the dropping of the atomic bomb. Einstein himself recognized this when he learned the news. When his secretary Helen Dukas told him what she had just learned from the radio that the bomb had been dropped, he said but one word—"oj weh" (woe is me). It is the only time in his life that he used Yiddish as an expression from his heart. He knew at this moment that his life was shattered. Later on, he would tell Linus Pauling "I made one mistake in my life." This mistake had sprung forth from his foreseeing the Holocaust years before it happened when no one else did. In 1939, he signed the ominous letter to President Roosevelt which he later recognized as his single mistake. His life was in shreds and he only remained a shadow of the person he had formerly been because he knew something had gone awfully wrong with his life and with the world. He was maybe the first person in history who connected the fate of everything with his own past activities. The bomb is of course a permanent risk for the planet, and the planet has indeed almost undergone a nuclear war since the moment "the physicists learned to sin" as Oppenheimer called it. He had been responsible for building the bomb in the first place and thereafter refused to participate in building

A. Sanayei and O. E. Rössler, *Chaotic Harmony*,
Emergence, Complexity and Computation 11, DOI: 10.1007/978-3-319-06781-0_17,

the next bigger bomb. So there is this tragic shadow looming over Einstein's life after 1945 darkening his last decade. Up to that time, he had had the greatest fun in science, maybe greater than anyone else has ever experienced, because he was able to see new things that looked absurd to everyone else but which to defend was his joy, to thereby change the way everyone else on the planet would be thinking about physics. In this respect he was akin to Newton who also had been a very lonely mind. There even were insights which they both singlehandedly shared—like weightlessness when you are in free fall. This was the biggest insight Einstein believed he had ever had. When you jump out of the window, then while you are falling everything falling with you is just an inertial system—there is no force that could generate any acceleration any more—just like this holds true in empty outer space as everyone is familiar with today from live reports from the space station. Newton had had the same idea, as Thibault Damour dug out (as mentioned). Einstein in addition could rely on the newly discovered special theory of relativity. Poincaré and Lorentz had already seen special relativity—virtually everything except for two points which Einstein added. But they had not seen the "ontological dimension" of the main implication of special relativity—the fact that everyone in constant motion is at the center of a frame that includes anything that is not in motion relative to him.

The fact, implicit in Maxwell's equations, that c—the speed of light—is a universal constant was first recognized by Einstein as being centered on a single observer called Einstein, for example. He only once confided to an audience—in 1922 when he had just arrived by ship in Japan after learning the telegraphic news that he had won the Nobel Prize of the previous year—how it all had come about. His expression "Dank Dir—thanks to you—I have solved the problem" is contained in a talk presented in German to his Japanese audience. It was translated live into Japanese and from there into English. He had said this toward his friend Michele Besso, after having returned to his house in late morning after a long nightly discussion they had had in May 1905. Besso was his best elder friend of long standing and his colleague at the Swiss patent office. The outcome of the translation was "Thank you, I found the solution." This obscured the real meaning—that he owed the breakthrough to Michele. So Besso is co-responsible for Einstein's deepest first insight. It means that the speed of light c is a universal constant for everyone as if he or she were the center of the world. Hence if someone else is moving relative to him, the speed of light is the same constant around the latter, too—which is a contradiction-in-terms. How can the speed of light be the same around two people who each are in relative motion to the other? Einstein apparently had discussed with Michele Besso the following riddle; If Einstein was standing somewhere in the dark and a flash bulb would go off at his feet while at this very moment the fastest runner of antiquity, Achilles, was passing by: the flash would create an expanding sphere of light around Einstein but the same flash would also generate an expanding sphere of light around the fast moving Achilles, co-moving with him. Two expanding spheres that are not identical but both generated by the same flash of light seemed to represent a logical impossibility. Then in the night following the intense discussion with Besso, Einstein saw the solution in a flash; it is possible to solve this conundrum—it is not absurd.

The solution is that you first switch over in your mind to a simplification in which the sphere is no longer an expanding sphere in three dimensions but only an expanding circle in two dimensions. That is, you can omit one dimension in describing the situation. Then you can draw a picture in three-space, plot a vertical line in the middle with time going up around the non-moving Einstein. You now can understand everything pictorially. The expanding light sphere is now simply an expanding circle emanating around Einstein and expanding upwards in what is called a "light cone." In the same simultaneity of Einstein, we also have a slanted (no longer vertical) line along which Achilles is moving within the same light cone. Around Achilles, there now also exists an expanding light cone—the same light cone. Two different light cones would be in contradiction. Is it possible to have two expanding circles or perhaps ellipses within the same light cone? The answer is, of course! Einstein is standing, the light cone is expanding around him; Achilles is moving, the same light cone is expanding around him. But the expanding light sphere around Achilles now is not a circle at its upper end but rather an ellipse. What did we win with having one common light cone? It turns out that within the same light cone, Achilles also stands in the middle with an expanding light circle around him, while for him, now Einstein is receding backwards, with a slanted ellipse around him in the same light cone. That is, there are now two cuts through the light cone, each of the same standing. Each expanding circle is horizontal in the respective person's frame but slanted in the other person's frame. The price to pay for this synthesis is that simultaneity—what is simultaneous with each in his own frame—has become personalized. In this way, everything had become transparent and simple.

The equations of the Poincaré-Lorentz transformation were well known at the time. Poincaré had already seen the light cone, but the "ontological component"—that Achilles and Einstein are really democratic equals having exactly the same status as observers—had not been seen in a flash by anyone before. That night following the long discussion with Michele marked the beginning of modern physics. A new type of thinking had entered with Einstein who was not afraid of accepting insights which would have appeared absurd and disquieting to everyone else. What does it mean that each person defines a motion system, a frame, in which the laws of nature are the same, one for Einstein, one for Achilles? Does this not mean that nature acquires a solipsistic aspect to it? This fact notwithstanding, we can combine the two "solipsisms" into one picture. The observer was suddenly given a role that no one had thought of or accepted before: an absolutely privileged position. Relativity implies absoluteness.

How old was Einstein at the time?

That was in 1905 when he was 26 years old.

What was his motivation to work on special relativity?

Both Poincaré and Einstein knew all the equations. But no one understood them intuitively at first. This picture of the light cone with the slanted cuts was known in principle. It became a foreign type of thinking when taken as the central fact. It was not the formulas, they were known. It was the reduction towards a fact of life—simultaneity having become private—that Einstein embraced so as if he had come from another planet. Einstein as mentioned also saw two previously overlooked little things that had escaped the elder Poincaré's attention. The first was the so-called "transversal Doppler effect." If something is moving across you, then when you look at it while it crosses your perpendicular line of sight, you are bound to measure a redshift. This had not been seen before, but it is if you wish a very minor point. The other new point is again "ontological", the twin paradox. Einstein still called it the twin-clocks paradox; his friend Langevin would call it the twin paradox 7 years later (it does not matter whether the clock is an ordinary clock or a biochemical clock).

Is it really a paradox?

These paradoxes are only seeming paradoxes; the one with Achilles and the one with the twin, but they are both absolutely brain busting. Up to this day, high-school students experience grave difficulty understanding how it can be that if you move away and come back, you are younger than your twin who stood put. It is possible to draw a very simple diagram showing this. It involves *three* twins. Number one stays put, number two departs from the same position at time zero to the right (say), in order to suddenly veer back at a certain marked distance. Number three is originally at twice the distance of the second's return point. Twin number three starts approaching at constant speed at the same moment the second starts moving away at the same numerical speed. So, the second and the third meet (passing by each other) in the middle, both having the same time on their watches. When the third reaches the first, the second reaches the previous starting position of the third. So we get two symmetrically overlapping right triangles with the same baseline. Clearly the second need not veer back at the moment of reaching the middle, since he has exactly the same eigentime as the third at every moment, including the moment where they meet. The returned second and the third can stand in for each other. While they meet, they can exchange clock times, but this is redundant since these times are identical anyhow. This example (which was apparently given by Einstein in similar form in the 1920s once) does without mid-way acceleration. One sees that the moving third twin—like the second but without acceleration—has a slower-ticking clock the whole time. Hence he arrives being younger at the standing first twin on the left. Q.e.d.

The paradox is usually told with only two twins, the one who stays put and the zigzag traveler. This is irreparably confusing since a sudden acceleration enters which goes beyond the applicability of the special theory of relativity. Now the familiar fact that motion slows down the clock rate in special relativity suffices to prove the twin's paradox.

Using three twins, a triple?

Triplets, yes. So the twin paradox becomes a trivial implication of special relativity. Einstein always had much difficulty later explaining this to lay-audiences. He came up with the idea of two crossing space ships and while they crossed they should exchange their clock times. But the example with the three twins is even simpler, one feels. It clearly represents an unprecedented violation of common sense; that just by its being made to move, a clock should "ontologically" tick slower. One can, of course, slow-down clocks by cooling them. Then everybody will understand that the cooled clock is ticking slower compared to its non-cooled twin clock. That without cooling, just by moving, you should be able to change the rate of a clock without this fact being detectable to the people in the moving frame of the clock is a similar "scandal" as the story of Achilles. The twins paradox is totally absurd. Up to this day, it represents one of the reasons why the number of physics students is declining across the planet because the young shrink away in horror from such "obvious nonsense." Einstein is a miracle.

I think the main reason is the teachers making students hate the clock paradox.

I was about to say that good teachers—I know a good one—are forced by their students to doubt this result. Only then is a good relationship between the students and their teacher going to be maintained, as I found out. It goes without saying that this is a tragic development. Only by facing the courage in Einstein's spirit can one progress. The twin paradox is typical of the way Einstein's mind worked. He possessed a thoroughly provocative character if you wish. He was not afraid of finding things that everyone else in their sound mind would feel forced to reject in order not to go crazy. He fascinated his private students he had coached to make a living in the time he was jobless. They soon became friends and had no longer to pay him. The price to pay was that he brainwashed them. Those relationships turned into lifelong friendships, as if he had been a religious master. To be "scandal-prone" in science in the same way is very rare. No one else ever had this strength. Newton came close, but I would not know of anyone else, maybe Galileo or Giordano Bruno.

Einstein once said, "I have no special talents. I am only passionately curious."

He was not afraid of thinking. That had been one of the accusation points raised against Galileo: "To this man, even thinking is fun!" Einstein went on finding things that were "impossible." His third surprise insight—that mass and kinetic energy are inter-convertible—also from the same miraculous year 1905, would be the one that was going to shatter the world eventually since it made the bomb possible via the mind of Lise Meitner who thought of it at the right moment. The fourth "characterless claim" came in 1907 when he looked at the accelerating long rocketship in outer space. No one had ever seen a rocketship at that time. He was only thinking of a high tower on earth at first, and then replaced the tower by a constantly accelerating rocketship in outer space—an equally long object with $1g$ acceleration. He saw with his mind or rather with his guts that you can replace gravity—which no one understands—with acceleration. The acceleration must be valid across an equally long stretch. The accelerating long rocket-ship then internally mimics gravity. The advantage is that this seemingly unnecessary roundabout way (the imagined contraption in outer space) can be totally understood using the results of special relativity. There indeed arises no problem in understanding everything inside this artificial scenario. The hypothesis invoked by Einstein was that, maybe, what holds true in the imagined rocket-ship holds true also in the tower under gravity. Then, the rest of his life was essentially devoted to deriving the implications of this mental visualization. It then came a little bit as a catastrophe that this (fouth) "unwashed idea in special relativity" proved extremely difficult to fully picture. As fate had it in 1907, 2 years in the wake of special relativity, the fourth idea of the accelerating rocket-ship made him stumble into a trap that was unavoidable in the absence of quantum mechanics. One tiny little result that he hated when first confronted with (and that could only be repaired much later by quantum mechanics) was the reason that he abandoned thinking and publishing about gravitation for almost 4 years. It was his first drawback, the first time that the pure joy that imbued Einstein's mind—he called it his "black soul"—was a little bit troubled. Of course he was no longer so young at age 28. He felt he had to accept that something that had brought him so much fun—the finding that the speed of light was a universal constant—suddenly turned out to be no longer a universal but only a local constant even though he had put it in as an axiom in the first place. A first *non-sequitur* in his edifice. He felt that inside the accelerating rocketship mentioned, he had to reluctantly accept this conclusion (as expressed on the last two pages of his long 1907 paper on special relativity). I should perhaps briefly describe what it was that he found because only then we can understand the new solution. One can look down inside the brimming rocketship from the tip, and one can also look up from the bottom. You can then use your mind and the findings of special relativity. Essentially you only need your own mind since everything is so simple, in principle. Never before had a "gedanken experiment" been taken so seriously in physics. May I continue?

Please, do.

You can send a light beam up from the bottom of the constantly accelerating long rocketship towards the tip, and the same thing you can of course do on earth inside the long tower. This vision actually was implemented experimentally half a century later by Pound and Rebka in 1960 at Harvard, where they used the extremely sensitive Mössbauer effect of quantum mechanics to confirm Einstein's "gravitational redshift" as he had called his fourth major "absurd" prediction. The experiment proved that he had been right, and to date, everyone knows that all the clocks downstairs here on earth are slower-ticking than their twins up in the satellites of the GPS, in obeisance to Einstein's baton. This fourth purely mental result of the young Einstein was again the fruit of his incredibly disciplined spatial thinking. If you look at a single light pulse, or a photon, that is emitted at the bottom and is now ascending inside the long rocket ship in order to arrive upstairs at the tip, something happens to this light pulse inside the rocketship. He realized in his 1907 review of special relativity that the light that arrives at the tip ticks more slowly than it did locally when it was sent up from the bottom. The reason is—while the light is ascending inside the extended long rocketship, the ship is picking up further speed. Hence the tip of the rocketship has already acquired a certain forward velocity relative to the bottom when the photon arrives. Therefore, the light source at the bottom is effectively receding relative to an observer at the tip. This "falling-back velocity" is constant. So the arriving light is characterized by a constant redshift, that is, its frequency is reduced (the name is due to the fact that red light has a longer wavelength than blue light).

This is a very simple pictorial thought, but it was maximally hard to believe because it totally contradicts everyday intuition. I mean, the length of the rocketship is constant, so how can the bottom be moving away from the tip? This patent office clerk believed so strongly in his own way of geometric thinking, even in the face of the counterintuitive new theory of special relativity as shaped by Poincaré and Lorentz that he was sure that he arrived at a totally transparent picture again. Here he stumbled on an absurd new prediction—gravitational redshift! It represents the most astounding finding in his life and perhaps in the history of physics. The new impasse: the speed of light appeared to be no longer constant inside the rocket ship. It would remain constant in every local frame inside, but it would not remain so across heights. This is what he thought he had to reluctantly accept that in the lateral direction if watched from above, light would appear reduced in its speed by the redshift factor like any other process going on down there. That was a big tragedy to him because this had been the holy assumption from which he had started out in the first place—that the speed of light is universally constant. Now it suddenly turned out that in gravity, the speed of light is no longer universally constant—only locally which is much less. Actually, this impression of his—that the speed of light has to be sacrificed as a globally valid constant in order to understand gravitation—was unnecessary in reality as one can see to date (with the help of quantum mechanics as mentioned). So for once in his lifetime, he was not optimistic enough. He had been the most optimistic scientist up to this time and had been rewarded with unheard-of new consistencies. His optimism got blunted a little bit when he felt he had to give up the very insight that had enabled him to understand Achilles and the flashlight, and the transversal Doppler effect and the twin paradox and mc-squared.

Now I pull quantum mechanics out of the hat as I said, and everybody is ready to think that I am uttering a sacrilege when claiming that the impossible indeed holds true—that the speed of light is rehabilitated in its global constancy even inside the accelerating rocket-ship and by implication a tower on earth and elsewhere. I can do this only if there is a new implication to present holding true inside the accelerating rocketship. This point stems from quantum mechanics as announced. Quantum mechanics implies that photons that have a lower energy downstairs, due to their longer temporal wavelength (redshift) possess a proportionally longer spatial extension (wavelength), too. This could conceivably be locally masked. However, quantum mechanics also implies that atoms produced out of these lower-energy photons, which have a lower mass-energy (without this fact being locally manifest, just like the reduced ticking rate and increased wavelength), are as a consequence proportionally enlarged. For, they are created by photons that have a lower energy due to the gravitational redshift. The technical name is "positronium creation and annihilation." This quantum effect entails that another quantum effect applies: the Bohr radius formula of full-fledged quantum mechanics. The two quantum laws taken together imply that all masses downstairs are proportionally increased in linear size. This chain of two quantum effects which both lay in the future was impossible to divine by Einstein in 1907. Quantum mechanics therefore brings back a *global c* to special relativity in the equivalence principle (which looks natural in retrospect) but in consequence, also to general relativity. So far, no one has found fault with this continuation of Einstein's happiest thought. A doctoral student of mine who had found a proof of the new gravitational size change using angular-momentum conservation (root of the subsequent Olemach theorem) got a PhD denied 7 years ago. The present renaissance of Einstein's "*global c*" could not be foreseen at the time. The student now deserves rehabilitation much like Louis de Broglie, the student of Einstein's friend Langevin.

We know that Einstein provided major contributions to quantum mechanics in the first era of quantum theory on energy quantization. Can we talk about these contributions for a moment before continuing with other related points?

Einstein discovered in 1905 that Max Planck had found something very deep which Planck himself did not understand at the time, the explanation of the photo effect by introducing "photons." It earned Einstein the friendship of Max Planck from a distance which helped build his career after a delay of 3 years. Einstein had a different way of thinking than almost anyone else. On the one hand, he was very sharp in mathematics and could put much sense into equations, but on the other hand—and in the first place I believe—he had a very strong geometric intuition. Algebra is the most powerful when you first know in your mind pictorially what you are going to derive, then you can translate it back into reality. The pictorial mind can be no less sharp than algebra, despite the fact that it is "only" working with pictures. Descartes was also one of these doubly-gifted people—maybe they were the only major thinkers who had this "mutually potentiating" gift of being both maximally productive in geometric thinking and able to make the connection to freshly conceived algebraic equations. The notion of a "function," in a plotted diagram as every TV viewer is familiar with had first to be invented by Descartes. One could now ask; which of the two is more important, the equations or the visualization? I would say that if you can think in an exact way geometrically, you are done already, because this can then be translated into equations eventually. But I know there are powerful dissidents—like Wolfgang Pauli perhaps. Geometry itself was founded in ancient Greece on very simple drawings made in the sand. Pythagoras' theorem was already known to the Babylonians. So the connection between algebra and geometry is very old. Soon after, in Greece, Anaxagoras discovered the transfinite exactness, as you know. At that time, everyone was bound to use spatial intuition in the first place. The modern two-tiered development started more or less with Descartes, although Galileo and Kepler were forerunners. There is this beautiful historical book by John Freely. Descartes' "analytical geometry", to write down equations and simultaneously be absolutely sure that every step works for purely geometric reasons, was the historical breakthrough. Although this was a major development, the source of it all—geometric intuition—is in my mind still the root of science. One has to return to this root—as Einstein did—if one wants to make major progress. This is a confession of belief on my part. Maybe many people will disagree with this viewpoint. Nonetheless all the modern developments including the computer show that the marriage between both cultures is essential—with the leading part taken by intuition, not by blindly following rules. The most recent "Virtual Realities" manage to give the mind a new exact spatial universe to roam about freely in, as in *Neuromancer* (a 1984 novel by William Gibson).

Maxwell had a rationalization. He distinguished between two types of scientific thinkers: Some like himself preferred to climb mountains on their own in lonely solitude to get a panoramic view no one else had ever had, to thereafter awaken others to building a tunnel from one point here to one point there. Most people would thereafter prefer being tunnel builders and tunnel users, because you can arrive very fast at new results in this fashion. However, without the previous overview from the top of the mountain, this particular fast algorithmic route could not have been built. I see a connection between Maxwell and Einstein in this regard. Einstein was very much motivated by Maxwell's electromagnetic theory of light which had already led to the Michelson-Morley experiment. Maxwell himself had proposed the experiment in a letter published after his death. There is a direct chain from Maxwell to Michelson via a US military person if I remember correctly. So the influence of Maxwell on relativity theory is even greater than commonly appreciated (he foresaw it before his untimely death); it was not just his equations—which were clumsy until Heaviside reduced them from 18 to 4, as I recently learned. Maxwell's equations were then re-spatialized by Einstein's mind and in this way they enabled him to feel absolutely secure about how light is behaving. Poincaré had had similar insights—he exerted a strong influence on Einstein. Nevertheless Einstein was independent in his own spatial intuition. It is very unusual that intuition is given such a high ranking in science—in this case by two people in a row, Poincaré and Einstein. But let me return to your question.

Let me now ask you about the contributions by Einstein to quantum mechanics. You started out from Max Planck .

Einstein saw the first major implications of quantization which Planck himself had not yet seen. Einstein's paper on the photo effect is quite famous—I never studied it in detail. Einstein is here part of a larger if still fledgling community, so that he is not as towering here as in his own field of relativity, in which he progressed essentially alone. Einstein also had much influence on the quantum domain in the second decade of the 20th century—on the laser, for example. He had a lot of beautiful insights—even the machines which pilots rely on in their airplanes not to lose orientation (gyroscopes) are based on patents obtained by him. He also held a patent on refrigerators—many interests indeed. I think that his relativistic breakthrough in quantum mechanics was the biggest contribution which he made in 1935 with Podolsky and Rosen…

Apparently, he always maintained that his most revolutionary idea was in quantum mechanics. He did not look at his contribution to special relativity as revolutionary. Today, most physicists see it differently.

It still could have been a little bit tongue-in-cheek since everybody knew he was famous with relativity theory.

Do you consider his early photoelectric effect as being more important, or the much later EPR result? The latter suggests that an interpretation accepted by the whole scientific community—that of quantum mechanics—is not complete. But this dissent only concerned interpretation. An interpretation can in principle be changed later. One of the fundamentals of quantum mechanics, starting out with the photoelectric effect, stems from Einstein as we saw. The groundwork for quantum mechanics was laid by him. Was his deepest contribution perhaps the EPR paper?

One can say so. His geometric intuition (which was so maximally powerful in special relativity and then in developing the general theory) again showed up in this magnificent contribution. He saw a direct connection to the quantum riddle itself and insisted on predicting that eventually, quantum mechanics would become completely transparent. This is indeed the case in principle with endophysics where one tries to do exactly that, although no major progress was made for 15 years. This "lower-level deterministic theory" is very hard to sell and is actually very intimidating. When Einstein first spotted the perfectly "equal rights" that lie behind the different "frames" of special relativity, this event meant a big turn-about in science. I would provocatively call it "the religious turn-about" in physics. Let me explain this by returning to Einstein in Michele Besso's house in early 1905. In their discussion, Achilles was passing by Einstein at high speed while the light flash was emanating from their momentarily coinciding feet. The question had been whether the two light spheres or light circles, the one expanding around the stationary Einstein and the one around the by-flying Achilles, could be made part of the same light cone since both had to possess symmetric rights if c was to be a universal constant for both. We re-lived this already. This "visualization" was the basic breakthrough which Einstein made compared to Poincaré, who was the greater mathematician.

We were in Michele Besso's house already before.

I am headed for a new point. Poincaré as far as I know, never accepted the total equivalence of the two frames but would have preferred one frame—that of the universe—to be privileged over a totally observer-centered physics, which is what Einstein had hit upon. If special relativity is taken at face value, each person is at the center of an equal-rights version of the universe. If there is a way to make a legitimate transformation between these two equal-rights versions of physics, then we are confronted with a strange—solipsism-like—relativization of physics. Einstein was maximally sensitive to the reproach of solipsism. In 1920 he was publicly accused of adhering to a solipsistic theory of physics by Ernst Gehrke in a public talk. He was much more annoyed by this claim than by Gehrke's express anti-Semitism, and published a rejoinder in the "Berliner Tageblatt" on August 27 the same year. He was the most irritated by the claim that he had introduced solipsism into physics. Einstein is the one who dared come closest to this point in the history of physics. Relativity actually means being at the center of it all yourself, so it is the opposite of a relativization but gives the observer an almost "absolute" status. It is this experience of Einstein—that the very name of relativity which had been coined by Poincaré and was accepted by him—touches on the heart of physics.

The new meaning of relativity was that the observer is "absolute"; but how can two different observers be absolute each? It turned out there is a transformation, as we saw. Thereafter, there is nothing solid in the old sense left. Nevertheless there exists something that remains absolute, namely subjective experience—the colors, the Now, everything tangible in our own private experience. How can science reconcile this solipsistic stance in the world, which each of us has, with its objective description of everything that is going on if Einstein is right? The battle entailed by his combining the two views had been on Einstein's mind since early on, from the beginning of relativity theory and earlier, right back to his first notes written at age 16. He was, I would say, able to return in 1935 to check if, and prove that, quantum mechanics can be conquered if you add to it this relativistic way of thinking. He was accustomed to harboring frightening thoughts for years. It was his conviction that everything can eventually be reduced to his so hard-won discovery of equal rights for every observer. Or what is essentially the same, is the fact that the observer is at the center of physics in a way which is both totally democratic and totally privileged (Lévinas is a kindred mind here). On the one hand, we have solipsism, on the other, we have total democratic equivalence of everyone. These two views, which appear like absolute opposites to most everyone, were in Einstein's mind forced together into a new synthesis. This had been so rewarding an experience that he set out to apply this unique synthesis also to quantum mechanics. He felt perfectly sure that this relativistic type of thinking would also be able to solve the riddle of quantum mechanics.

Why did he never mention the relativistic viewpoint applied to quantum mechanics in the EPR paper?

I think it is because this was a trivial matter to him. It was absolutely self-evident to him that everyone would see that this is implicit in what he proposed. Only his followers would not see it because they were not Einstein and could not depart from a secure level of everyday thinking. As we saw he had brought in just one basic new idea; that instead of having one quantum object on which to make measurements, it is possible to utilize a twin copy. You can make two mechanical toy locomotives leave from the same standing position in two opposite directions as we have seen…

Actually, his first idea expressed in the EPR paper was more abstract. I think that David Bohm was the first to exemplify Einstein's idea using the neutral Pi meson at rest decaying into an electron-positron pair. But in the original EPR paper, the authors consider only two subsystems, A and B, and then look at an interaction or non-interaction being present between them. Einstein never proposed such an example or something like that, the paper was more general. Bohm added some elements—so that people sometimes speak of the Einstein-Podolski-Rosen-Bohm experiment. The latter is only one possible interpretation of the EPR idea.

Thank you and if an idea is not completely finished, it is not as clear-cut as it becomes when people look at it from a later position, as David Bohm did. Einstein's "naughty" idea was that instead of having one quantum object, you can have two. I am sure he was aware of the old high-school textbook example of the two equal toy locomotives of zero-sum momentum, releasable into two mutually departing symmetric motions by unleashing a spring that had been compressed between them. Twenty years later, in 1955, my high-school professor, Dr. German, demonstrated it, and the textbook he used was most certainly much older since he was grey-haired at the time. Unlike me you studied the paper from the quantum point of view, and your dissenting with my simplistic view is very helpful. In my view it is this symmetry which allows one to violate the uncertainty principle by making a combined momentum and position measurement on two independent but "correlated" particles.

I do not understand one thing here. The Einstein idea somehow contains the concept of correlations between these two subsystems. But it was clarified later by Bell that these correlations only mean that we have two probability distributions. The word "correlation" *only has a meaning if we have distributions, but the notion of a distribution goes back to a statistical viewpoint on quantum mechanics, and Einstein did not agree to this very viewpoint.*

I would not debate that he believed in this statistical viewpoint, but I lean to the interpretation that he did not use it in his mind because he was convinced it can be explained.

But it did not lead him to a statistical interpretation of quantum mechanics.

Yes, he believed in the statistical interpretation only under the condition that in reality, there lies a deterministic explanation behind it. So he did not contradict the formalism itself, he only wanted to explain the formalism in a causal fashion.

Max Born believed that quantum mechanics is essentially statistical. Einstein did not agree with his friend on this point. In several letters to Max Born, he says so. On the other hand his idea about EPR presupposes the concept of "correlation." What is the meaning of correlation? Does it mean that we have two distributions that are correlated?

He did not think in terms of equal distributions, I would think. He believed in identities of two parts.

Maybe he just wanted to show that we can simultaneously measure position and momentum on one particle…

Exactly, that by using two twin particles, we can measure at leisure the momentum on the one and the position on the other. Then we know both features on both, that was the idea.

This sounds like David Bohm's example. The original idea of EPR is in four pages, and is a very abstract thing. They did not mention any example.

Maybe, because there were two coworkers, he was not allowed to tick so simply. In the background—and in the title of the paper which is often forgotten to take into regard ("Can quantum-mechanical description of physical reality be considered complete"—that is, can quantum mechanics perhaps be completed?)—, there is the idea that you may exploit having two copies of a quantum object. He renounced of adding that instead of having two "spatially separated" ways of accessing two correlated particles with correlated quantum features, you can of course choose two "causally separated" ways of accessing their joint quantum features. This is what the word completion ("complete") in the title of the paper expresses.

Do you want to say that special relativity somehow existed here in the background?

Precisely. To him, it was self-evident and trivial that you could add special relativity. He believed the idea was so strong already that it was not really necessary to add special relativity. This is because he had already convinced himself that you could complete quantum mechanics by measuring two non-commuting variables—like measuring the position here and the momentum there, on both sides independently. It would be a miracle if your measurement here would have a faster-than-light influence over there. Also this would contradict special relativity. What John Bell discovered later was that you could formulate an inequality proving that the first measurement does influence the other in a superluminal fashion. This indeed proves to be the experimental outcome, and is very hard to understand directly, even though no information can be superluminally transmitted in this fashion as it was later shown.

Do you consider the reaction of Niels Bohr after the EPR paper to be satisfactory or not?

Certainly not. I looked at it once. He missed the point. Everybody missed the point at the time, namely, that Einstein had genuinely tried to "kill" Bohr's theory with this paper because "completing" quantum mechanics by measuring both position and momentum violates the uncertainty principle and especially the complementarity principle that Bohr had erected as a corollary or if you wish, smoke screen.

It seems that the main idea was contributed by Heisenberg and Pauli to the Copenhagen viewpoint.

Yes, and with them the majority of physicists stood behind Bohr. Bohr realized that the article was a direct assault on him, on his own scientific identity.

Another question and maybe it is none of my business to bring it up, but from a historical perspective it is interesting to ask; why did some other highly known people in quantum mechanics never talk about EPR or write something against or in favour of Einstein's idea?

I would say it had to do with a deeper understanding since everyone else had misunderstood Einstein's EPR paper in terms of the power of "aggression" against the established view that stood behind it. Namely, the fact that Einstein really wanted to "kill" quantum mechanics in the interpretation which had come up at this point in time.

But he did not suggest any novel interpretation. Even if one supposes that Bohr's interpretation is not satisfactory, Einstein still did not suggest any new interpretation. He just wanted to falsify Copenhagen?

He wanted to show that the theory as a whole is untenable. That it is basically flawed, since the word "completion" is very contentious.

Is it not necessary to offer a new theory in order to allow science to continue, otherwise after the old theory is gone we would have nothing left in our hands?

But you see, killing a theory …

Is also important …

Is the necessary first step, and only then can one hope to find a new theory. This is nothing one has in one's own hands; one cannot know beforehand how the shreds of a refuted theory are going to look.

Do you believe that EPR was the main motivation for Everett to formulate his new theory of quantum mechanics?

I feel tempted to think so. But no, let me take this back. The EPR paradox was only one of the things that Everett explained in his paper. He mentioned the EPR paper in his second short dissertation that got printed in Review of Modern Physics in 1957. Almost no reader saw that on the last page, he explained the EPR paradox in terms of his own theory. Apparently, Everett was very much influenced by a letter written to him by Einstein when he was 12 years old (as already mentioned). This fact made him so strong. To return to Einstein, the main point of his and Podolsky and Rosen's article was that one could make use of the good experience he had had with special relativity (keyword "spatial separation") and look at correlated twin quantum objects, to thereby find a violation of quantum mechanics. This is what actually was done later. As far as I know, the idea was taken fully seriously for the first time by Susan J. Feingold in a paper which was never published, apparently, but was mentioned by her thesis adviser Asher Peres in a joint paper of 1978, namely, that you can understand Einstein better by adding special relativity to the EPR chapter. This to Einstein himself was implicit and self-evident, but everyone else has been blind to this. Once you start to think in this direction, it turns out that you can actually fulfill Einstein's promise that it is possible to kill the Bohr version of quantum mechanics, as the "relativistic EPR experiment" shows. Specifically, if the two measurements are made in two different frames which are mutually receding, this causal separation makes sure that you now have two Copenhagen-type experiments in one. In the one frame, you make the measurement first in the momentarily receding satellite, and in the second frame first down on earth. In the other frame, on earth, it is the other way round ("VX diagram"). Then you have two sets of quantum measurements that are each independent from the other since in each, the wave function was reduced first. Both measurements will be entangled by showing the Bell correlations. This causes a contradiction. If you believe in only one quantum world existing, then something is broken. What is broken is the formalism of quantum mechanics itself because we now have these two independent measurements of position and momentum (in fact their analogs). Nevertheless everything is fine with quantum mechanics regarding each pair of measurements if the one frame is assumed to be earlier and the other later. If one frame is the first and the other the second regarding the measurement of the non-commuting observables recorded, everything is absolutely fine. It then does not matter on which side you are. However, if both sides are privileged in terms of making the first measurement, then both position and momentum have been measured independently and Copenhagen is dead. This was apparently Einstein's intention. On the other hand, we are not finished in this case. For in the other—later—interpretation, Everett's, the measurements are not described by the same double result. In this case we have *two* pairs. What the second group on the ground records, and gets relayed by the space-bound first group, is not the same as what the space-bound group records and gets relayed by the earth group. Here we have 4 measurements in effect. Of those 4, only two are accessible to every individual observer, in a way which is perfectly compatible with the formalism of quantum mechanics. There is no longer "too much" information for anyone. The people on the ground are happy with the result because they have made the first measurement, with the later one in the spaceship being Bell correlated just as it should be. The people in the space ship are happy, because they have made the first measurement, with the subsequent measurement on the ground as relayed to them showing the standard Bell correlations as well. The appalling thing is that this "peace" carries a price-tag. While Copenhagen is killed, what survives is Everett's objectively observer-centered quantum world. While the Copenhagen interpretation is refuted, the Everett theory is confirmed and becomes the only theory of quantum mechanics left. The same proposal was made independently by Roger Penrose in his book "The Emperor's New Mind" in a drawing on page 287. Nevertheless no one seems to know the fact that quantum mechanics as Einstein knew it at the time has been effectively killed by Einstein. The experiment is in preparation—only the quantum satellite is still to be finished, as I learned from Anton Zeilinger, as mentioned.

But instead of saying that the whole theory has been killed, it is perhaps more appropriate to say that we now need another new interpretation?

Exactly. After Bohr's theory has been vanquished by Einstein, it is the theory of Einstein's former young correspondence partner, Everett which alone survives in quantum mechanics. Everett's interpretation hence becomes Everett's "theory" of quantum mechanics. This is very hard to swallow, though, because it has this solipsistic streak to it that Einstein was so maximally afraid of in his youth, until he could show that both of his two special-relativistic observers, the fast Achilles and the standing Einstein, could be fitted into one light cone, being just in different "cuts." This time now, there are two mutually opaque whole worlds. It is very hard to figure out what Einstein would say to this scenario in the footsteps of his former protégé, Everett.

In every world, the result of both measurements is fixed. So we are not allowed to have a simultaneous measurement of momentum and position in either world?

This is the solution, but we have to pay the price of more than one quantum world existing. We again brush by solipsism. It is not quite as bad as it looks because we still have the unchanged overall wave function. It is only "projections" that are mutually incompatible and inaccessible. Still, John Bell's friend, the Dalai Lama, would come into the play. Each individual then lives in his or her own quantum world. Bell depicted the problem by making the Everett worlds successive rather than simultaneous. This would be a version of the immortality of the Dalai Lama who lives in a different body every 90 years or so, without consciously remembering the previous existence. This would now happen on a micro-fine time sale, but with the same mutual opacity. This was Bell's point in the paper "Quantum Mechanics for Cosmologists." From every second to the next, we would live in a different world, not just in a different Now. We are anyhow at the mercy of the Now-change in the world. The Now, too, is solipsistic, which fact is not very well known (although Roger Shepard offered a proof almost 20 years ago). We know we are being propelled through the Now as if riding on a chair lift. I was once riding through a ski resort in a lift where each passenger was confined to a little seat of his own, being tied to the seat and with about 10 m distance to the next seat (in which my wife was riding) and with a deep abyss below each. I felt very lonely and it was a very intimidating experience, much more intimidating than riding in an airplane. This loneliness of the rider of a ski lift is an apt illustration of how the world is experienced by us at every moment in the Now. This solipsistic type of thinking, which can also be called a religious type of thinking, was palpable behind Einstein's special theory of relativity and it is now again behind this EPR theory, it appears to me. Einstein was afraid of it as we saw, but he had the courage to face it. This attitude is even more impressive than if he had loved the idea and had presented it in a boasting way. He was afraid of his own idea, but he was sure that somehow the riddle of quantum mechanics must and can be solved. Even if the price to pay was to be called a solipsist eventually, he took the risk. He did not like what he said himself but he insisted on a rational solution.

Thank you, we now can close our session on Einstein in which we mostly talked about the EPR today. Thank you very much for your time.

Chapter 18
Intelligence, Orangutans and Second Darwinism

We already talked about the brain equation in detail, and we made the connection to dynamical systems and biology. It may be a good idea to talk about intelligences. What is human intelligence? And what about intelligence in animals? What does it mean if one animal is smarter than another? Can you make a link between the brain equation and this topic?

Let me try. There is a book by Willie Smits which I can mention at the beginning. He is a forest scientist originally. Many years ago he was given a dying orangutan baby in Borneo into his hand. He found it on a heap of trash. This baby looked at him so convincingly as only a little baby can, he started to adopt it. Then he became the big friend of orangutans in Borneo and is now one of the leading experts in Orangutanology. He is the co-author of a 2008 book titled "Thinkers of the Jungle." It has a provocative title and presents many results which are completely unknown to the scientific community about the behavior of orangutans. For example, they love to see photographs and they remember exactly what the photograph represents of many years ago when it was taken. Of course they can use mirrors very freely. The name "orangutan" itself means forest person or forest people. There is a story which goes like this: Why do you call them forest people? The answer: Because they do not speak. And why don't they speak? Because they know that in that case they would have to join the labor force. So it was believed that their higher intelligence prevents them from speaking. This reflects a great respect for these forest people. Science largely neglected having a close look at these—as I likewise believe—most highly developed primates of the planet including humans. However, the brain weight is only about half the human brain weight. So according to this criterion, they should be less smart. On the other hand, we know that magpies and some other birds with their tiny brains exhibit comparable intellectual capabilities not just to lower primates but even to apes, so brain size itself is no reliable criterion. If you live in the trees, there is a selection pressure for reducing weight wherever possible, so the brain weight of orangutans is probably influenced by this fact. But that is just a very indirect argument which proves nothing. I do have an argument though which can prove that they are more intelligent even though nonspeaking than most other species on the planet. These

A. Sanayei and O. E. Rössler, *Chaotic Harmony*,
Emergence, Complexity and Computation 11, DOI: 10.1007/978-3-319-06781-0_18,

slow and cautious animals—or people maybe—have the genetic trait that they produce only very few offspring during their life-time. A species can only continue if a female member produces at least one viable successor during one generation. Otherwise the whole population would die out. If the procreation rate is very low, this means that the efficiency of child care and of making sure that the youngster survives to adulthood must be very high, since otherwise this species would no longer exist. One can now look at the numbers; an orangutan female can start having a baby not before the age of 8. Then there will be another baby only every 8 years because they only raise a single child very individually and with very much attention paid to it in close bodily contact over 7 years.

Biologically can they have more than one baby during the 8 years or not? Is it a decision made by them not to have a baby?

Since they are treating their baby with so much care for 8 years, they do not get an oestrus, they do not become ready to conceive again for so many years. Therefore, an orangutan female who lives about 50 years under good conditions can only have at most 6—and in reality maybe 5 or 4 children during her life time; say, 4. Then two of them on average are female. So a female is reproducing with only about two female births per life. This means that at least one of two must survive in order to become the mother of the next generation. This is a maximally low number. For example, in humans procreation is much more efficient than in orangutans. As far as I know, there is no other animal species which has so low a reproductive rate. This is very surprising because these creatures do not live in a secure environment but in the tree tops. Every night they build a nest out of branches and leaves which they somehow put together successfully. The branches move in the winds, although there is not much wind fortunately in the forests of Borneo. It is very hard even to think of having to live under such precarious conditions as a human being, of being able to successfully build a safe bed for the night out of random tree branches and of finding the right food in different parts of the year, knowing which types of tree are carrying food at what time, which food is dangerous, which is helpful under what conditions, and so on. In spite of this highly demanding life style, they have this maximally low rate of accidents as we saw. How can this be? The only answer that comes to mind is intelligence.

Many years ago, I read a report that one had made a test with orangutans in comparison to humans in which the subjects were offered a choice between two objects that were both attractive. One was closer and less attractive; the other was farther away and more attractive. Human children would prefer the more attractive item even if it is farther away and more risky to take. The orangutans invariably chose the less valuable but safer object in such a situation. This could mean that their brains are tuned towards being cautious. They also are quite slow in most of their movements, although they can be also very fast, and by the way they are very strong, very much stronger than humans, which is no wonder. Their low procreation rate in a very demanding environment proves mathematically that their brains must be very efficient. The low procreation rate and the respect this demands was also emphasized by Willie Smits in the book mentioned. It actually represents a mathematical problem as we saw, posed by the procreation rate. So the efficiency of the brain which these creatures carry is proven to be singularly high developed. The question that poses itself is whether it is possible to rate brains according to their functional capability. Here "deductive biology" enters as I call it, the fact that in biology one can explain observations in terms of theory. A "Theoretical Biology" in the same sense as the familiar "Theoretical Physics" was pioneered by Ludwig von Bertalanffy whom I met, and by Nicolas Rashevsky, the teacher of my friend Bob Rosen. Theoretical biology is a mathematical discipline just as theoretical physics is. It tries to derive facts known from biology in terms of deductive reasoning. There is one example that everybody knows about, it is Darwinism. Darwin had to explain the fact that even though the environments in which the species live change over the course of time, species often survive even though they may at first be ill-adapted to a new environment. How can this happen? Darwin realized that it is mandatory to have endogenous changes which create an alteration in behavior or bodily features such that, in the changed environment, the altered species can again survive. This is a very general principle which cannot be denied. It is not very predictive in detail because it by definition depends on what chemical and biochemical and physiological features these organisms already possess. It thus depends crucially on history. The whole prehistory enters. Although Darwinism is a deductive theory, it is not very strongly predictive. It is predictive only for some general things. The most surprising prediction implicit in Darwinism is the existence of a "Brain in the Genome," as Michael Conrad and I called it in the 1970s. In the genome, not only is the whole evolutionary past still represented, but at the same time the genome is becoming more and more efficient in making mutations in the right domain as a kind of directed mutability. This is a controversial issue in theoretical biology. No matter how it is solved, however, the point is that the theory of Darwin represents a purely mathematical insight. It says that when a parameter is changing in a given environment so that the survival rate is affected, the biological system is bound to produce endogenous changes including one that overcomes the deterioration of survival. This is a mathematical argument. Amongst biologists, the fact is not very well known that Darwin's theory is a purely mathematical one. It is complex, it looks as fuzzy as biology itself looks taken as a whole, but at the same time it is a very strong, deep, provable insight—a theorem in other words.

In the same category, one can find a second fundamental theorem. It arose in a discussion with Konrad Lorenz who was 63 at the time. There is one phenomenon in evolution of which I had not taken much notice before—that there are animals living at the sea shore where the tides are coming and going in regular intervals and where the animals have to move away from the tide or to bury themselves when the waves come or to switch their metabolism in time. One already knows empirically that these animals possess built-in clocks which tell them to change their behavior just in time. This we already saw. Lorenz pointed to the fact that this information—of when to hide or when to do something else at the sea shore—cannot possibly come from the genes. Genetic information cannot tell you at what hour the next tide will come. But you can have a "clock" which can be synchronized by cues from the environment. Then you have information in time which is essential for survival, but you need something like a receptor to influence an endogenous oscillator to make sure that you can survive in such an environment varying on a time scale that is much shorter than a generation period—that is the difference. From this example we were then able to deduce that this second type of information, which is not genetic but is nevertheless essential for survival, is a new element that needs to be added to Darwinism. This new subfield does not suffer from the disadvantage we had before—that everything depends on history, on the pre-existing chemical properties of the organism in question. We realized that this is a type of information which is independent of how you are built. Virtually every organism living in this tidal realm will have to do the same thing—to build a clock and receptors in order to be able to predict what happens. This is a new temporal element in Darwinism. Survival depends not only on long-term temporal changes as Darwin had shown, it also depends on short-term temporal changes as this oscillator theory reveals. However, with this identified as a principle it had now also become possible to see that survival very often also depends on position in space. For example, look at a moth and a candle-flame. The moth is attracted to the light and in many cases the moth will die in the flame. But moths living in a region where the trees need fire to grow, like the beautiful giant sequoias in California, will eventually need to adapt to the fact that light can be dangerous and develop a special sensor—for example, a heat sensor. One thus realizes that if survival depends on short-term temporal or spatial conditions in the environment, then to take heed of these by means of special sensors does not depend on the chemical and biochemical evolutionary pre-history of the organism in question. Invariably, what is needed is to respond to some environmental cues in the right fashion.

Survival depends on much more predictable things here than was the case in the theoretical biology of biochemical evolution itself. The slow, long-term evolution follows different laws than this new problem of adapting to very short-term changes that cannot be predicted by long term selection, by virtue of being shorter than one generation period. This was the insight arrived at in 1966. It took me 6 years to realize that this is a problem which is actually well-known in mathematics but which no one apparently had connected to biology at the time, the so-called "travelling salesman problem." We talked about it before, but it is just too tempting to try to put it into a "rounded-up" form again if I may: You have a map with a number of towns indicated, and you have a travelling salesman who is selling something and has the task to visit these towns at minimum mileage costs. So he should use the shortest connecting line between the given points on the map. This is a simple mathematical problem and one thinks at first sight that it is trivial to solve. Everything consists of distances that are known. Nevertheless it turns out that if you increase the number of towns that have to be visited, the number of comparison-type length checks you have to make in order to find the shortest connecting path rises very sharply. The number of choices that are open grows exponentially, and the difficulty of solving the problem therefore likewise rises neither linearly nor polynomially with the number of towns, but rather exponentially or more. This problem is called an "NP-complete problem" where "NP" stands for "non-polynomial" which essentially means "exponential" rather than merely polynomial which is infinitely more harmless in a sense. The word "complete" means that this problem in its nontrivial simplicity stands for very many other problems which likewise are so embarrassingly difficult to solve although they look trivial. If one of them can be solved in polynomial time, all of them can be solved. This class of NP-complete problems now possesses a "sister class" in which it is not the shortest path that has to be chosen, as in the NP-complete travelling salesman problem, but in which a slightly different travelling problem needs to be solved. The traveller here has a car with a fixed number of different gas tanks. Each needs to be filled up in time before it runs empty, with a tank-type specific time interval or driving distance. The filling stations for the differently colored tanks are distributed in space independently. Yellow, red, green, blue comes to mind. The constraint here is that you must not miss one tank filling. As soon as the remaining travelling distance is shorter than the distance to the next tank filling station of the same color, then you will die, that is, the car will come to a standstill and you have lost the game. This problem turns out to be just as hard to solve as the original travelling-salesman problem. I called it "the travelling salesman-with-alarm-clocks problem." 6 years later it got rediscovered by Garey and Johnson under the name "decision-type travelling salesman problem." It belongs to Informatics, but even more so to Theoretical Biology in the hard Darwinian sense which parallels Theoretical Physics. This problem has to be solved, for example, also by orangutans. It has to be solved by all organisms travelling in space. For this purpose, they need a "brain." One can go so far as to say that "the function" of a brain is to solve this decision-type travelling salesman problem. Essentially the same problem was independently discovered by Eric L. Charnov in a less mathematical form under the name "optimal foraging theory" as a formal subdiscipline of the science of ecology. This is the existential problem of how to survive as an organism endowed with locomotion. Garey and Johnson proved that the decision-type travelling salesman problem is as demanding mathematically as that of the NP-complete ordinary travelling salesman problem. After this excursion, we can return to biology.

Every organism whose survival depends on position in space needs to solve the TSWAP (travelling salesman with alarm-clocks problem). Hereby the addition of more realistic conditions—unequal tank radiuses, different refilling durations, presence or not of reserve tanks, etc.—make no essential difference. The TSWAP problem therefore defines a deductive approach applicable to the BRAIN (also 5 letters). The job of the brain thereby turns out to be independent from biological history and to be purely mathematical. You actually arrive at a theory of brains in biology which gives you a whole series of functional designs as a function of the ratio of the mean distance between sources to the source-specific maximum travelling radius. Mathematics yields a bonanza in theoretical biology that almost rivals its role in physics. If the sources are much closer together on average than your travelling radius covers, very simple measures suffice to secure survival. There is a book by Konrad Lorenz of 1973 titled "Behind the Mirror" (which I like to quote because it quotes me—if you allow the joke) in which he gave the history of all related ecological and ethological approaches in the usual descriptive way of traditional biology. Many devoted people had compared different types of behavior in simple organisms and in more complicated organisms over centuries. For example, there are single-celled organisms which just travel around irregularly without sensors. Obviously, the resources available for food are so dense and easy to harvest in this case that it suffices not to stay at one place all the time. The next version of behavior was called "kinesis" long ago. It means that the organisms also travel irregularly, but with the difference that when they are close to a point of high concentration of nutrients, they just slow down due to the presence of a sensor. Here the speed depends on the environment, on where they are. This already allows them to survive under the given density of sources. Next, there are some worms which already have direction-sensitive sensors. Eventually, when the distances between sources (our filling stations) become larger relative to the maximal travelling radius, you need to know still more precisely where to go. Behavioral ecologists worked on this phenomenology for centuries. Ethology later fitted in. But the next steps on the ladder were less easy to spot from the comparative empirical point of view of the field researcher. If you do not go straight in the direction of a vitally needed source because the pertinent tank is running out, you will have lost if the remaining tank-filling radius is overstepped. One still finds a hierarchy of solutions known empirically to biologists for many decades, but previously not classified functionally. The deductive approach offered in the 1973 paper ("Adequate locomotion strategies") enables the hierarchy of more and more effective solutions to be continued. It implied a distinction between insect-type behavior (choice between a discrete set of alternative goal-directions) and spider-type, and finally the locust, mollusk and vertebrate type of optimizing behavior (locust because the mantis shrimps came in since). This classification had been impossible before in empirical biology. It is great fun to see that field research and mathematical deduction can converge, something which was previously possible only in physics. One realizes with perplexity that it is possible to formulate by deductive means a functional brain theory. The abstractly deduced solution involves CPU's (central processing units) combined with "motors" and "sensors" and eventually very good distance sensors called "eyes." In the hierarchy of functional brains found, an important threshold exists. The brain equation with its purely local optimization features represents a "plateau" which when reached does not yet include more sophisticated solutions which involve "supra-local" or what is the same thing, "path optimization." For this purpose, the design needs to be modified. The simplest modification is to leave it unchanged, but to add an additional device, namely, a so-called universal simulator. Then several local solutions, chosen from different, momentarily not actually held but simulationally adopted positions, can be elaborated by "locomotion without (overt) locomotion" and evaluated before the best one is chosen in the sense of being connected to the motors. However, this is of course only one possible way to solve the path optimization problem on hand. There are other strategies of global optimization possible. But let us first look more closely at this threshold-type solution where local simulation becomes insufficient. It represents what was called the brain equation. It is based on sensors having specified the positions of the closest sources or filling-stations for the differently colored tanks or clocks. It computes from those directions, and all intermediary ones in 360° (for animals living in a 2-D environment for simplicity), the best locally evaluated direction in which to move.

Finding the best local solution is a well posed mathematical problem—or rather sub-problem. It is possible to find a set of space- and time-dependent nonlinear additive functions for every direction. Spiders are not quite there, vertebrates etc. are. It hereby turns out that you need *two* such functions of the momentarily valid distances, of the closest sources of the given n types of tank and the momentary distances and directions of the pertinent closest sources identified with the on-board distance sensors. The one type of function is valid under "optimistic" assumptions, the other under "pessimistic" ones. The two are added with different time-dependent weights in dependence of the "sum state." Paradoxically, even if there are only "positive sources" (no predators or abysses), there are both attractive and repulsive directional weights involved. The temporal and spatial shape of these could be indicated. Both are nonlinear. The spatial ones involve the circle of Thales or a cardioid (for repulsive directions), respectively. The temporal ones are more complicated, involving rational functions of distance and time. Slight adjustments are no doubt still possible. One type of source—procreation-enabling ones—yield a special shape whose attractive positive component can in principle diverge. The outcome of all this is what I call the Brain Equation. It represents a time-dependent force-field which describes the attractiveness of all neighboring sources in the environment and, more specifically, of all directions in space including all those going backwards. For example, going in one direction will bring you closer to two sources with a modest angular distance present in front of you, on coming closer, you are drawn more strongly towards one of them. This is a local problem centered about your momentary position. The equation contains potential functions or rather their gradients. Every direction in space has an attractive and a repulsive component for every source type (color). When the gradient or the sum of all these potentials is taken, a locally optimal direction is determined. When you follow this locally optimal direction according to the brain equation, then you will have a fairly good survival record. But there is a limit to the efficiency of the brain equation. When the ratio between the travelling radius per source and the mean distance of sources of the same type is reduced still further, then the brain equation is no longer sufficient for securing survival. It then needs to be complemented by a "universal simulator" or "cognitive map machine" or synonymously "virtual reality" (VR). But let me return to the brain equation.

It can happen that one direction acquires an absolute priority (an infinite weight) if the travel is to continue. This absolute priority over all other directions means that you get an infinite gradient in the brain equation. That is, an infinitely strong force is generated internally by this mathematical machine. It is like a Newton-type equation with potentials, but here it is the gradient, not the level line, that is followed. We have solved the "travelling salesman problem with alarm clocks" (or synonymously the "decision-type travelling salesman problem") with an equation which is suboptimal. It is not a very highly intelligent equation because it yields only a locally optimal solution. When the local solution no longer suffices, it becomes necessary to evaluate several local solutions starting from differing fictitious points in the neighborhood as mentioned. You here put yourself into a position in space at which you are not at the actual moment, to evaluate the optimal direction valid from there, and so on from several places. One can then, as mentioned, add another machine of a much simpler mathematical structure which is a universal simulator or VR so that you can still use the brain equation. You would then in simulation go to another place and try from there, and then compare what you found, finding a potentially much more optimal solution than you did from the original local point of view. The combination of the brain equation with a universal simulator allows you, in principle to optimally solve the travelling salesman-with-alarm clocks problem. One can predict that biology "cannot not use" this principle. You could object that I might instead choose a chess-playing type of solution where everything is tried out in discrete steps with some other algorithm. That would also work. We know that chess-playing computers are very smart overtaking human players. But the chess-type solver is not as valuable as the brain equation combined with a universal simulator. This is because in reality, you sometimes do not have much time to "think"—that is, simulate. If this can happen, as is very realistic in biology, then you need the brain equation anyhow for a first trial in order to avoid doing something as nonsensical as just sitting there and thinking. If you have more time a moment later, you then can elaborate. So the brain equation offers a kind of emergency solution depending on how much time you have to think. If you have more time due to the first gut-type reaction, you can improve on your solution further by using the brain equation in the simulation mode. So the brain equation actually is a universal necessity in biology, and this not just in terrestrial biology, but in every kind of biology. It is totally independent of the biochemical or chemical background of a particular type of life. You can therefore say that "brain life" is a form of life that is independent from the chemical background of the organism in question. The brain equation is the same for all organisms wherever you are in biology or on Mars or another planet, wherever life exists in the cosmos, including even nuclear-chemical life on neutron stars as Robert Forward conceived it. They all have to use the brain equation. So the brain equation is very important. Forgive me if I got carried away…

I did not want to interrupt.

You are right; this is not where we started out from today. We started out with the bigger question of what is the difference between human beings and animals. The brain equation with its added universal simulator is the most intelligent machine one can think of in my opinion. One can prove that this machine including the simulator becomes as big as the moon or the sun or the solar system and eventually the whole known universe if it is to solve the most difficult and longest travelling salesman problem. Gödel thought about very difficult solutions in mathematics. The famous "Gödel incompleteness" means that there exist certain mathematical problems which he could prove are not solvable in finitely many steps. This travelling salesman-with-alarmclocks problem—just like the original travelling salesman problem—happens to yield a Gödel-hard problem in the limit. So the travelling salesman problem in its two versions has a connection to this deepest and most arcane problem of mathematics, the Gödel solution. It even turns out that one can now understand why this solution is Gödelian. Why is it that Gödel saw that it overtaxes the human mind to consider his problem which has no unique solution as he and his successors like Paul Cohen taken together proved? It is now possible to say why the Gödel solution is so strange. It is because it represents the limit to the travelling salesman problem. The latter problem makes the arcane Gödel incompleteness theory intuitive because the travelling-salesman problem is so close to our own survival at every moment in every decision. I did not specially mention that you need memory—long-term memory—although the universal simulator automatically contains a short-term memory for handling its differing conditionally simulated paths as we saw—if the environment is redundant, which is generally the case in biology. If the environment does not contain redundancies, you do not need long-term memory along with the brain equation. Long-term memory only involves a hard-ware type addition to the brain equation. But this memory is not very essential. A person with Alzheimer's disease is still able to save a child who fell into a river, for example. So memory is strangely not of the essence for a brain. But I realize that so far, we still did not yet talk about humans or the orangutan in his relationship to humans.

Maybe we can continue with this later. Perhaps you want to conclude by summarizing the fundamental things briefly?

Let me try my best. The orangutan as I said is very cautious. He is thinking a lot but he is not talking and he is also not using a sign language. So it is not the vocal organs that are missing which are responsible.

They do not make sounds?

They do have some calls.

There are also people who do not talk a lot.

And there exist human people who cannot hear and cannot produce sounds. So talking is not "very" important although every language is an infinitely rich and precious thing. Nonetheless these orangutans lack something else that humans have—maybe Willie Smits would contradict me here, but equally likely he would specify the difference himself. He speaks of the "Thinkers of the Jungle" which means that he has a very high opinion of the jungle people whom he is trying to save. Nevertheless everybody realizes that these beings are "only" apes. There is something missing with them even if it is true that their brains are more intelligent than the human brain is, hardware-wise. We know already that an orca brain, or the sperm whale brain which is the biggest and by far most complex brain on earth, may be more intelligent than the human brain, and similarly for the elephants, too, and there may also be a more intelligent species of mantis shrimps or octopuses—such as *Architeuthis* perhaps. So the question is; can it be that there is some fundamental difference between humans and the other highly brained animals? According to the universal brain equation, the latter no doubt are more highly developed in terms of the universal simulator employed, the hardware involved in thinking. But there is another criterion according to which humans apparently are "superior" (or at least radically different functionally) because humans can easily wipe out the remaining last thirty thousand or so specimens of orangutans of the planet, and humans can even wipe out the whole planet and reduce it to a 2 cm black hole—which is something that orangutans apparently are not capable of doing. How come? What is the decisive difference between orangutans and us? It has to do with the use made of the brain. But what is the difference between intelligent human beings and intelligent forest people? This is at first sight an unsolvable conundrum. It is a very delicate problem which can be posed now in the wake of Willie Smits' book, "Thinkers of the Jungle." I mean, people so far despair before this question, but it can be solved.

Chapter 19
Hubble Law, Cryodynamics, Cosmology and Technology

A new session. We again want to talk about cryodynamics, but this time on its applications. We talked about the theory in two previous sessions. My question is as follows; what is the application of cryodynamics to cosmology? What are the new results that we arrive at?

Cryodynamics grew out of cosmology—or rather it grew out of a result which is the essential result in cosmology, the Hubble law. Hubble's extrapolation was very courageous—not at all justified by the data points that he had in hand. He divined that the redshift observed with the few galaxies measured at the time (most done by Vesto M. Slipher during more than a decade) would be linearly distance-proportional and as such would continue over very large distances. This daring hypothesis would prove correct even though it was almost crazy to assume. People at first did not believe that a distance-proportional reduction of energy of photons in the universe could possibly hold true. One of the first who acknowledged this very possibility was Fritz Zwicky who worked not far from Hubble in California. Zwicky saw that you can actually derive this putative implication of Hubble's finding from first principles which made it much more trustworthy. He could anticipate the generality of the law because he envisioned a general physical principle that would actually predict the phenomenon. He predicted that the law was as global as it later turned out to be. However, it is not quite true that he was the first, Abbé Lemaitre had already had a similar idea, but it was less deep and was more guesswork, whereas Zwicky had a physical law to offer. Still it is true that Friedmann's expansion principle evoked by Lemaitre was likewise up to the task at the time. Zwicky did not introduce a new postulate, only a yet to be verified new law implicit in accepted facts. He conjectured that light traversing moving galaxies—which were known to exist at large distances due to Hubble's previous discovery—is subjected to interactions. He pictured that light traversing this cauldron of moving galaxies would energetically interact with galaxies. It was a strange and powerful abstraction from reality—to look at a galaxy as a point particle, and at a photon as a point particle, too, only both endowed with differing

A. Sanayei and O. E. Rössler, *Chaotic Harmony*,
Emergence, Complexity and Computation 11, DOI: 10.1007/978-3-319-06781-0_19,

mass. To anticipate a systematic interaction between the two was a mighty jump of the imagination that Zwicky introduced here. It lies in the tradition of the first statistical-mechanical theories which also were originally deterministic in the sense of chaos theory. The word "gas" itself means chaos.

Gas theory stems from an old school of thought which aims to understand the complex, out of very simple interactions between particles. This interaction would be Newtonian-deterministic in kind. This thinking is very idealistic in the spirit of a Poincaré-type celestical mechanics which today is called chaos theory. For some reason, Boltzmann and Poincaré never got together, but Boltzmann had very similar ideas himself. Zwicky's idea involves both Boltzmann's and Poincaré's type of thinking. Zwicky saw that the present interaction is not repulsive, as it is in statistical thermodynamics. If you introduce attractive forces, however, you suddenly have a new many-particle theory. Then the possibility arises, as Zwicky implicitly saw, that the interaction is not entropic but the opposite. That is, the heavy particles do not give away some of their energy in the interaction, dissipating it towards the energy-poor light particles that are passing through as would be the case with billiard balls of differing masses and energies. On the contrary, the slim fast particles give up part of their energy to the energy-rich fat slow particles. A very unusual sort of thinking which predictably met with immediate resistance on the part of the scientific community. I find it admirably powerful, showing that Zwicky was a very original thinker. He would underscore this with other discoveries that he made. He by the way was also the first not to be scared by the prospect of the sun becoming a red giant 5 billion years from now, when the earth will then be eaten by the sun, as everyone believes. Zwicky saw that one can move the earth away from the sun by some weak constantly acting machinery in the category of rocket boosters. Zwicky was able to conjure-up an easy solution where everybody thought there is none. He had the habit of provoking people by proposing things that appeared absolutely ridiculous at first sight. It is a good strategy in science, in physics in general, to force yourself to be impossible. He seemed to possess a recipe for that. But to now return to the redshift law, he conjured-up the possibility that all the beautiful laws of thermodynamics—its axioms and theorems—are limited in the sense that there exists a totally different class of interactions. Microscopically speaking, the latter would not generate thermodynamics, but an opposite science of the same caliber. This theory if it existed would rather deserve the name cryodynamics since cryós means cold in place of thermós, hot—an obviously absurd idea, attributed here to Zwicky.

One question; I think we cannot use the word "absurd" here because according to this Zwicky-type viewpoint, the interaction has been changed to being attractive. The word "complementary" comes to mind here.

Yes, cryodynamics would not be an absurd thermodynamics; it would only be so in the mind of people who think they have to stick to traditional thinking at all cost. Zwicky divined implicitly that attraction could entail such an unfamiliar consequence—that big guys, the rich people, get even richer at the expense of the poor guys, namely, that the energy-rich heavy particles would draw away energy from the energy-poor photons traversing the cosmos. This appears almost ridiculous at first sight. Sonnleitner calls this very principle “energetic capitalism.” How Zwicky could even have a hunch in this direction is a mystery. In our course on chaos theory in Tübingen, we came up with essentially the same idea when the century had just started. Zwicky got reinvented so to speak. We—that is, Dieter Fröhlich and Normann Kleiner and the rest of the crowd—started out with the question of what happens to light passing between randomly moving mirrors. We at first falsely thought, as mentioned, that in that case, the light would be decreased in its energy. We saw the connection to cosmology and wondered whether the Hubble redshift law might not be caused by an analogous interaction with moving galaxies. Eventually we corrected ourselves seeing that the moving mirrors, being repulsive, would generate the opposite effect—an increase in the light particles’ energy. But the attractive “mirrors” or galaxies would indeed predict the Hubble redshift. So we independently more than 70 years later re-arrived at Zwicky’s conjecture as a provable result. Nevertheless we for many years did not know that Zwicky had envisioned the same thing already. Then we learned from Ramis Movassagh that Chandrasekhar built a bridge, yet without knowing that he supported Zwicky, with his idea of “dynamical friction”—of heavy stars braking very fast ones. While this still overlapped with ordinary statistical mechanics, where very fast particles are also retrieved back into the fold by a kind of braking force, it obviously was the same idea again in principle. Chandrasekhar was struck by the same idea as Zwicky. Even though this idea contradicts the laws of thermodynamics, it is—if correct—a part of statistical mechanics, too. Or more properly speaking, it is a part of deterministic statistical mechanics or, even more properly, of deterministic chaos theory. Chaos theory thus potentially describes the Hubble redshift law. In the next step, one then has to check whether it does so only qualitatively or whether it does so quantitatively.

Then my second question; what is the application of cryodynamics to particle physics? You already mentioned the ITER project.

If it is true that we have here a whole theory of the same caliber as thermodynamics, namely cryodynamics, then this is a big new field in physics which is bound to have applications of a practical type down here on earth. There exists a major technological problem with a concrete many-particle system, the so-called ITER in Southern France. Using the ITER, one plans to harvest the energy which fuels the sun also down here on earth, by enabling nuclear fusion to occur between different types of nuclei under controlled conditions. The ITER would produce energy in a much more clean fashion than the usual nuclear energy sources do. It would solve all of humankind’s energy problems—if this machine could be brought to functioning overcritically within a few years’ time. It had been hoped to start working in the last decade, but this dream did not bear out. Now the start has been postponed by two or more decades. Many billions of dollars have been invested already. So the program has come to a virtual standstill in spite of being pursued very actively. Also, the analogous project in the United States which uses not a homogeneous gas but pellets shot at concentrically with fast particles, does not work so far. In Texas and Japan, there exist related projects, maybe also in China. This is one of the big projects of humankind. What is the reason that it does not reach the threshold? The reason preventing this in the hot-plasma can be indicated. The Tokamak torus magnetically confines the hot plasma so that it can self-heat; generating new energy through nuclear fusion as in the Sun. However, the plasma with the nuclear interactions unfortunately also generates unpredictable turbulences as a magneto-hydrodynamic system. The main reason why there is no chance that it will work any time soon is that the hot plasma is dynamically unstable in the magnetic confinement that has to be used. When the hot plasma reaches the walls of the Tokamak in a turbulent protuberance, the toroidal metallic container at the moment of contact immediately cools down the plasma locally by many orders of magnitude, causing the whole process to break down. No one has so far found a way to make sure that this "breaking-out" can be avoided, even though the beginning turbulence can be detected locally in time in principle.

What is your solution?

The proposed solution is that one could locally cool the too hot region before it actually reaches the wall. That is, pushing the ghost back into the bottle locally. Such a local cooling mechanism was unfathomable so far—you could not cool by shooting in particles up until now. Even the idea looks absurd, but cryodynamics says exactly this; you can cool with hotter particles. You only need to concentrically shoot in even hotter particles. Galaxies have more kinetic energy than the photons from which they take away motion energy. What kind of more energetic particles could one locally inject into the hot plasma so as to cool the hot nuclei there?

You want to shoot some particles in? Why should the interaction between these particles and those inside the plasma be attractive rather than repulsive? They need to be attractive in order for the theory to work?

Yes, in order for the theory to be applicable. If you can find attractive particles to shoot-in concentrically with sufficiently dirigible nozzles, like electrons, this could cool the too hot protons and thus solve the problem.

Should you shoot-in hot electrons or cold electrons?

The electrons must be even hotter than the locally too hot protons, which is paradoxical at first sight. The even hotter electrons would take away energy from the too hot protons if injected concentrically at the hot spot to create an even hotter bath there. If you can direct several beams of very hot electrons to any point in the Tokamak, which would not be a big technical problem in principle, you could cool locally in time. You would need tiny holes in the walls to shoot-in through…

But there is a theoretical problem here: You explained the whole theory in classical physics, but now you are talking about an application which is also governed by quantum mechanics.

I am not so sure that this matters much. The temperatures are so high that you can safely apply classical thinking to the plasma particles, including Maxwell's theory, of course.

But you are shooting-in somehow very cool particles?

On the contrary, hot particles need to be shot in, even hotter than the ultra-hot plasma itself, hotter by a factor of one hundred, perhaps. The whole idea looks ridiculous—that when shooting-in even hotter particles, these should not increase the temperature locally but rather decrease it. This is absolutely absurd at first sight.

I have a question here. Suppose someone carries out this experiment and the result is not the one that you predict: Can one then conclude that the theory you offer is falsified, or not?

Good question.

Do you consider this experiment as a way to falsify your theory?

As a potential application of the theory, it would be falsified. However, applications are always very much more complicated than the pure case that a theory offers. There are natural hot electrons inside the plasma. So everything is quite complicated since there will be interactions with the local electrons. Everything depends strongly on the involved densities, and on hydrodynamic processes. So there are many possible reasons why the idea as a whole might turn out to not work. Nonetheless at the present moment in time, there appears to be a realistic chance that it will work.

Should you not first consider a simpler problem? For example, we could set up a tiny system of hot protons and shoot hotter electrons in, both being less hot than in the ITER. Then we could see whether we can indeed have such a cooling.

I thought of this. For example, there is a famous experiment from the early 20th century where you have electrons circling in a vacuum, in a not quite perfect vacuum, and you can watch the electrons circling in a magnetic field as a green ring because they emit some light. This beautiful experiment was invented by Walter Kaufmann in the first years of the 20th century. You can tune the energy of the circling Kaufmann electrons. There was a proposal that I once arrived at jointly with Jürgen Parisi as to how one could interfere with the speed of the Kaufmann electrons in order to verify cryodynamics empirically. My personal attitude would indeed be that you do not really need do that. It is easier to verify cryodynamics in numerical simulations. The 2-particle simulations that we did in a T-tube are examples. This simple system is very hard to run numerically. It is presently impossible to continue with these numerical experiments since funding agencies are so slow. In this simulation the numerical accuracy proves to be of paramount importance. As soon as the accuracy is reduced, the phenomenon disappears for numerical reasons which are not fully elucidated. On the other hand, there exist very many multi-particle simulations that have been done in the past and are still performed in galactic dynamics—how galaxies form out of stars and so on. Even the Big Bang is being simulated in studies involving millions of simulated particles (although no expansion exists if cryodynamics is real). All these beautiful computer simulations never revealed this mathematical phenomenon of cooling by hotter particles—only the two-particles T-tube simulation of Sonnleitner was successful so far.

How about quantum mechanics? Can you predict whether the theory works there, too, or does it not?

It should. Quantum mechanics is absolutely compatible with cryodynamics, just as it is with thermodynamics. There must be a regime in cryodynamics in which quantum effects take over and presumably cause a drastic modification. It will be a very interesting field, just as important as we have it now with low-temperature thermodynamic experiments.

As a last question; can you consider an area of application for cryodynamics in biological physics?

Not at the time being, I am afraid. I feel this is not very probable because of the fairly low temperatures in biology. But this brings me to thinking in the opposite direction—the subnuclear domain. There exists the possibility-in-principle that cryodynamics has an analog or even application in subnuclear physics, that is, in the domain of the strong force. This appears highly unlikely but the incredibly fast time scales and huge energies involved do admit analogs to thermodynamics to be encountered here, although I never looked into this. So we could leave this here as an open question; is it possible that apart from thermodynamics-like phenomena, we also have cryodynamics-like phenomena in elementary particle physics?

Chapter 20
Taffy-Puller, Topology and General Relativity

There is a paper by Jim Yorke. He tried to find a mathematical model of the taffy pulling machines as a chaotic system. I am interested in your ideas concerning taffy as an example of chaos. Can we see hyperchaos in this example? And what is the importance of this physical model? Is it only a plaything or does it have consequences in chaos theory?

Thank you. I encountered the machine in 1976 and saw that it has importance as a model of chaos. Later I got a letter from Art Winfree who pointed me to it independently. I had featured the taffy puller in my chaos-in-education chapter in Hungary with several pictures. The taffy puller is a wonderful example of a chaos-producing machine involving stretching and folding in a repetitive form as in a cross section through an attractor. It looks miraculous if you can watch such a machine in a sweets shop. Everyone would expect if asked that some taffy should fall off and that the whole thing would disintegrate, but it does not, it works for hours and months on a stretch. It is an example of chaos. You can think of two radiating molecules or two little light bulbs placed into the taffy. Then you could film the two light points moving inside the taffy in the dark. Many theorems of mathematics are illustrated in this analog implementation of chaos theory. It is a very nice gadget. You cannot have hyperchaos in the taffy puller because you only stretch and fold in one direction. You would need a 2-dimensional taffy puller in which you implement the folded-towel mechanism. Then you have two positive exponents, two directions of exponential divergence. The two expanding movements taken together produce hyperchaos. As far as I know, such a machine is not being used in technology, but I would not be surprised if some examples were to be used in the future. The folding-and-stretching technique is very old and is for example used in the production of Japanese steel swords, where the metal is being heated and then hammered out toward twice the width and then folded over, to be heated again and so forth. Then all inhomogenities including air bubbles are eliminated from the metal in this mixing process, which is self-stabilizing. The sword is also a nice illustration of a Poincaré map.

A. Sanayei and O. E. Rössler, *Chaotic Harmony*,
Emergence, Complexity and Computation 11, DOI: 10.1007/978-3-319-06781-0_20,

Nature is continuous, it represents a flow, but you can look at a cross section through the flow which then is a Poincaré map. The taffy puller is the simplest explicit illustration of a chaos-producing process in nature. You can take taffy or just play-dough and think of all types of repetitive operations that one could do with it in reality. The human mind of adult people is not flexible enough to conjure-up a catalog of all possible types of basic operations that one could employ. I once predicted in late 1976 that there exists an attractor which I called an "X-attractor" in 4-dimensional flows, which has not been identified so far. It would possess a three-dimensional cross section which one could implement with play-dough in ordinary 3-space. So this is an open question; what is the next qualitatively new phenomenon beyond chaos which arises in four dimensions? There is no doubt that one can find it with 3-dimensional taffy and/or play-dough. Every recurrent continuous motion in four dimensions can be described by a cross section which has three dimensions. The latter can be visualized as a 3-D map which in turn can be implemented by taffy, by play-dough.

The taffy-puller makes use only of a one-dimensional stretching, but you can do much more with play-dough. Nevertheless it is very hard to think of new ways to do something repetitive with play-dough. I remember I got two pieces of play-dough as a present on my third birthday, one was white and one was brown—two round slabs. And I began to handle them, to do things with them, eventually it was a mess. I mean, I had blobs of brown play-dough and I had pieces of white play-dough, and I wanted to mix them, to fuse them. So I took little pieces and put them together in all possible ways, to no avail. It did not occur to me that I could stretch the whole thing and fold it and stretch. Maybe it was a little bit too hard to do it easily, but the very idea did not occur to me. I still feel ashamed that at the age of three I did not invent the taffy puller which is a perfect mixing machine. It is very easy if you have two pieces of play-dough, one white, one brown, to mix them perfectly: Just put one above the other, pull them to twice the length, fold over, pull, fold over, and after 30 steps, I guess, you would have a light-brown mixture which shows no signs anymore of the original constituents. It would be a molecular mixture. I was too stupid to find the stretching mechanism, but I now know that there are other mechanisms that you could use to manipulate repetitively 3-dimensional play-dough. The minds of all mathematicians of the planet has not been strong enough, as yet, to find something new here that is typical of the fourth dimension in flows, making itself felt in three-dimensional space. It will not just be a one-dimensional pulling, or a two-dimensional pulling. Presumably it has to do with cranking, with twisting, with turning-around. There must be a radically new mechanism of how to both pull and twist, in order to get a new prototype attractor. I once proposed a project tongue-in-cheek 35 years ago: to spend a million dollars for equipment and for many volunteers who would go to the kindergartens and give play-dough to children and film what they are doing with it. One of these children among 100 or 1,000 or more will come up with this new mathematical operation which the mind of mathematicians has not been able to identify so far. This is the "play-dough big science" research program which no one ever took seriously. I am sure that progress in differential geometry depends on this program—including understanding general relativity in a deeper fashion. It all depends on a child making this invention and being filmed while doing it. Grown-up mathematicians have too many blockades in their minds to find the very simple solution. As soon as one sees it, one will realize what it is. It would then determine a flow in four dimensions as a radically new phenomenon, as important in four dimensions as chaos is in three, as oscillation and periodicity is in two dimensions and bi-stability is in one dimension. So there is a hierarchy. We have so far only conquered three dimensions with chaos, the fourth is still open and the fifth as well. This is a mathematical challenge in differential topology—I believe.

Why do you believe that progress in relativity and differential geometry depends on such a breakthrough?

Chaos is a differential topological phenomenon which became famous to my great amazement in the 1970s after Ed Lorenz had found the first popular example in the 1960s. Steve Smale in the 1960s and other people in the 1940s had already come across it and before them, of course, Gaston Julia and Henri Poincaré. The whole thing comes from the mind of the inventor of topology, Poincaré. He was a very rare spirit who influenced the whole of mathematics. Think only of his invention of the seemingly absolutely stupid idea of topology, of not looking at quantitative details but just at the few things that topologists are interested in, the number of holes and twists.

You are envisioning a bridge between chaos theory and general relativity?

Yes, certainly.

Could you explain a bit more? What is the bridge between chaos and relativity?

Poincaré and Einstein are the two names here. Differential topology was invented by Poincaré and Einstein invented...

...*General relativity* ...

...as an important element, not of differential topology but of differential geometry. The two are closely related but these two minds, these two fields, never got fused. The topologists believe that they are more high-ranking than geometers, and the geometers maintain they have the highest-rank topic and pretend they never need topology. So there is a jealousy in general relativity against the chaos people. The former say they are the pupils of Einstein and the latter say they are the pupils of Poincaré. It is of course cute, the whole thing, but differential topology or topology as a whole is standing behind differential geometry. It is almost a historical catastrophe that the connection between differential topology and differential geometry has not yet been made, only because of the jealousy between the French followers of Poincaré and the German followers of Einstein. Poincaré is not as well-known as Einstein is, even though both are very similar in their thinking. Poincaré came very close to what Einstein found later. While Einstein, formally speaking, found only one or two little new things in what

Poincaré had described before and had already called "relativity theory." Einstein also found something maximally big which Poincaré had overlooked. It is like the two directions in the not yet existing two-dimensional taffy puller. They cannot be reduced to each other. Some basic elements of general relativity were also already in Poincaré's book "The Value of Science" published by Flammarion in 1904 and discussed by Einstein with his friends in the "Olympian Academy" which they had founded. Poincaré describes there that the size of objects might depend on their position in space and shows how strange the world would become under this condition. Olafur Eliasson, the famous mathematical artist, later implemented an approximation using two light sources of different colors in orthogonal positions and a landscape made optically opaque by artificial fog. He is the same Eliasson who suspended a running electrical fan from a long rope and wondered what would happen (a beautiful hyperchaotic motion).

But to come back, Poincaré and Einstein were kindred minds. The "lights" that Einstein could spot within special relativity—especially the ontological twin paradox, and the person-centeredness of simultaneity, a whole private umbrella around every individual that stretches out to infinity—Poincaré was too old to appreciate. The impish streak in Einstein—that he always said he had a black soul because he drank black coffee—was inaccessible to the more seriously playing Poincaré—although topology and homoclinicity have their rebellious streaks, too. Since Poincaré died so early (in 1912), they could not become friends in time.

The question is; there are two fields here and apparently, their bases are similar, topology and general relativity. But to my knowledge, we have not heard so far that there exists a reasonable bridge between them to go back and forth. In particular, I have not seen anything to the effect that progress in chaos theory could help general relativity or vice versa. Can you envision such a situation? What help could chaos theory offer in general relativity? And what is the help that general relativity can give to chaos theory?

General relativity suffers from being a little bit too dogmatic and narrow in the tools that it uses. It essentially only uses curvature so far, it never talks about stretching as being of equal rank—the taffy-pulling experience is missing, if you wish. There is some stretching implicit if you look at Flamm's paraboloid, for example, which is the traditional geometry behind the Einstein-Rosen bridge, but stretching is marginal compared to the free reign that is given to curvature as governing the structure of differentiable manifolds.

The wormhole as a paraboloid?

Yes, the wormhole, this rotated paraboloid of Flamm, is the best illustration. It looks very beautiful, like a symmetric vase.

A bridge somehow.

It has a big opening up here and an equally big opening down there, and in the middle it has a narrow feminine waist. This is the wormhole. Einstein-Rosen-bridge is the right name for it. It not only contains curvature, it also contains stretching. However, this stretching is very limited. In general relativity, stretching does not play the same role as curving. It is somehow suppressed in the subconscious of the minds of general relativists for almost a century by now. They seem not to know that tearing is just as important as folding, pardon me, curving. The stretchability of taffy has yet to be brought in.

Why do you think that stretching is very important in general relativity?

It is a bit premature if I utter this guess, but I would dare say that in general relativity, if you do it right, you have to include just as much stretching as you have folding-over, that is, curving. I see here a historical fault in the reception of Einstein's ideas, in the sense that people overlooked the second topological implication of nonlinearity in general relativity. They would have avoided this blindness if Poincaré's ideas had been included as they deserve. So the whole field, rather than opening up, became a little bit dogmatic as a subfield of differential geometry which can here be characterized simplistically as "differential geometry without stretching"—there is some stretching, but it is under-represented. So general relativity can be "healed" as it were by learning from differential topology that stretching, which is the basis of chaos theory, is just as important in understanding space-time geometry as curving is. Actually, the two go hand in hand. It is very easy; whenever you have infinite curvature, you also have infinite stretching.

It seems when such a progress is made, the dogma will be overcome that there was a big difference between French and German scientists and a productive friendship will arise at last.

It is the European reunification in mathematics, one could call it. The whole world would profit from Poincaré and Einstein becoming friends in retrospect.

One could start like this: One could write a letter-type publication, for example, to suggest that the people working in differential geometry and general relativity should start working together on the potential contributions of "stretching" to general relativity. But apparently the people at the moment rather work on deepening the gap in between the two schools. I expect and hope that this gap can be made to disappear somehow.

Thank you for saying that.

Chapter 21
Sinai Billiard and Plane-Tree Alley

In this session I would like to come back to cryodynamics again. Let me ask you to explain it from the basis of the T-tube model, its structure and conception. In addition, I plan to ask you one question; I saw a paper outside physics of 2008 by Professor Zadeh, titled "Is there a need for fuzzy logic?" After more than 40 years of work he confronted this question and tried to answer it. The question came from a professor of electrical engineering who had put the whole fuzzy logic into question in that paper. Zadeh's paper was a whole-hearted response. Let me ask you the same question here: Is there any need for cryodynamics? And if so, what is it?

The shortest answer would be that chaos theory demands this new implication—cryodynamics—to be taken seriously, for, the whole thing just stems from chaos theory. Chaos in this light becomes even more fundamental than appreciated previously. We know that celestial mechanics is applied chaos theory, since chaos is an implication of deterministic Newtonian dynamics and of Einsteinian dynamics. Only quantum mechanics causes some problems here, but these problems do also show that chaos theory must be underlying quantum mechanics as well, which fact may come as a surprise. At any rate, the case of attractive fast moving particles interacting belongs to chaos theory from the beginning. At the same time there is no reason to criticize a statistical theory, but chaos is more fundamental. I am thinking here of deterministic systems, of constructing a many-particle theory of mutually attracting bodies. Hamiltonian celestial mechanics includes such a many-particle theory, just like Hamiltonian statistical thermodynamics does. Only that we in the new case obtain a statistical cryodynamics—the opposite of a statistical thermodynamics. (Thermós means warm and kryós means cold in Greek). Both sciences have the same rank, but the attractive case was not recognized as a field of its own with the same rank and birth right as enjoyed by the repulsive case of thermodynamics. It is indeed very unlikely, a priori speaking, that a big old discipline like thermodynamics should possess a sister discipline which is no less big and just as old, but somehow got overlooked.

A. Sanayei and O. E. Rössler, *Chaotic Harmony*,
Emergence, Complexity and Computation 11, DOI: 10.1007/978-3-319-06781-0_21,

Can I ask a question for clarity? What kind of many-body phenomena do we have in physics that cannot be explained using statistical thermodynamics, so that we need to invoke this new discipline?

It is the interaction of many particles that are governed by attractive rather than repulsive potentials. In other words, it is a theory of many-particle celestial mechanics. Celestial mechanics automatically contains such a subdiscipline dealing with many gravitationally interacting particles. When we look at the night sky, this is cryodynamics in action. Cryodynamics is a very old field in terms of its constituent elements and interactions—it just did not have a name before. The statistical or many-body part of celestial mechanics got overlooked as a problem in the history of physics, ever since 1687. This is almost unbelievable. Cryodynamics has always existed as a white area on the map of physics, but since it was white, people were misled into thinking there was nothing there.

So you have another starting point to explain cryodynamics than the T-tube model?

Thank you for this question. Just last night, I was thinking about the simplest possible model, and the idea occurred to me that I could draw a connection between Sinai's chaos theory on the one hand and cryodynamics on the other. Sinai, so I should add, is the inventor of modern gas theory in the context of chaos theory. In 1970, he published a very long and big paper on "a tennis game in an orchard," as a friend of mine, Harry Thomas, called it, as mentioned. Harry was a professor of theoretical physics at the University of Basel in Switzerland. We had many discussions when I was very fresh into chaos as it were. He had returned from a visit to Moscow and told me about people discussing the dripping faucet there, thinking that the irregular falling of the droplets was caused by air molecules interfering with an otherwise periodic droplet formation. He already knew about chaos, and we agreed that this was a deterministic phenomenon, even though almost no one believed so at the time—that the irregularly dripping faucet was just an illustration to Poincaré's theory.

To come back to our topic, he told me first-hand how one could interpret Sinai's gas theory. Sinai had invented this wonderful deterministic model for a gas of many mutually repulsive particles—an almost hopeless case to understand deterministically. It occurred to him to reduce the repulsive balls in three-D toward repulsive round disks in two-D, on a plane. Since this was still too hard, he saw that one could just look at two. And then fix the one disk by nailing it to the middle of a quadratic billiard table. Then the other would do all the work without loss of generality as it were. Further, the moving disk could be shrunk to a point if the stationary one's diameter was doubled, with no ensuing change of trajectory. Yes, then he identified, without loss of generality again, pairwise-opposing sides of the quadratic table. Can you visualize this? This is Harry Thomas' "tennis game in an orchard." An orchard is many trees planted in regular intervals. Then we can have a ball flying straight through the orchard without friction. You just need to look at the trees as regularly spaced repulsive disks, with the ball following an uninterrupted course. This is because the single cell with the repulsive disk in the middle, repeats itself indefinitely on an infinite tessellation space as one calls such a grid. (Like the advertisement for "Wissoll chocolate" from the 1950's: "Infinitely good!"). If you make the orchard a two-dimensional tessellation space, you have Sinai's problem.

As a billiard model?

Yes, as a Sinai billiard. It is a great help to have this orchard before your eyes, to have these regularly planted trees and the ball flying straight in between them. Sinai was thereby able to make a convincing connection between deterministic dynamics in chaos theory and the qualitative behavior of many-particle systems; the old problem of statistical thermodynamics. He was the first to look at many-particle chaos in a deterministic fundamental way. This picture of a tennis ball flying through an orchard gave me a mental association last night. I remembered a certain alley in Tübingen called the "Plane-tree Alley", Platanenallee in German. This alley I know from my early childhood days. I remember walking there when I was 2 years and 4 months old, making a classical remark to my mother. I was walking there with my mother and my little sister of 9 months' age in a carriage…

You mean the alley in between the two arms of the Neckar river in Tübingen?

Exactly. It is a long artificial island in the Neckar river, and the tall trees are very regularly planted and beautiful. The trees actually are leaning slightly outwards, away from the middle line, presumably having more light on their foliage this way. The direction of the alley is approximately west-east. When you walk there you pass by the "Hölderlin tower" as a half-round protrusion on the northern side of the Neckar river, where the allegedly insane late poet lived for more than three decades. He would on his long walks, lift his hat before every dog he encountered but not before any person. He had put hopes in Napoleon before, it is rumored. I was walking there with my mother and my little sister in her carriage. A little man, a midget (Lilliputian) in size, was coming across us. When he was within listening distance, I was asking my mother in an excited loud voice "Is this man already 9 months old?" He broke into a loud laughter and my mother would explain to him that my little sister was just nine months old. It was a good experience I had in this alley. Now I thought of this alley again in the context of Sinai's ball flying in an orchard. In Sinai's case the ball is not assumed to fly through in exact straight parallelism in which case it will never touch a tree but forever remain on its straight course. This conceptual model of many trees in a row and a ball flying through is very interesting also if you replace Sinai's "hard" repulsion by a "smooth" repulsion. What does this minor modification mean for Sinai's result? We still have the orchard, we still have the tennis ball, but now we have smooth repulsion. The result then, of course remains basically the same. If the ball is slightly non-parallel, we encounter chaos again. We have but one new phenomenon; it is no longer hard-disks chaos but smooth-disks chaos. The latter is more complicated because it involves Komogorov-Arnold-Moser "KAM" islands. Nevertheless we can still look at these trees and our unstable situation when the path looked at runs right through the middle of the tessellation alley. The trees now have a smoothly repulsive force acting on you, the ball, running in the middle. You come closer to two trees as you approach the middle line in between them, but you never touch a tree. So, all that occurs is that from time to time you have to negotiate the slight elevation in the middle between two trees, a saddle point in the parlance of differential geometry. That is all. It is a mental game like a computer game. But now, make the trees move—this is the new idea. We keep the symmetric path in the middle of the alley or orchard, but make the trees move, all equally and symmetrically. We can make them approach periodically, all of them doing the same thing. Or equivalently, but easier to visualize, we can let the trees stand and make them "breathe"—become broader and narrower periodically. Then the path in the middle will be affected somehow. It will be interesting to see what happens.

The path would be narrowed and widened?

The path does not change in its course, but the speed along the path is now no longer strictly periodically modulated in time. We now have a second change compared to the hard Sinai case; the speed does not just vary periodically any more, as it did when the smoothened trees did not oscillate. There is now the added possibility of a systematic, progressive effect on the speed. This is a new element, there is the possibility that the Sinai game…

He had fixed disks.

Yes. If we fix the positions of the smoothly rising trees, then there will be only a periodic speed change, as the ball musters the saddle point in between the two mounds that lead up to the tall tree stems and form a saddle point in between them, as we saw. But if we now allow the trees to move periodically and symmetrically without touching (or let them breathe in width periodically for simplicity), we do have a recurrent influence exerted on the travelling ball as it passes symmetrically in between the breathing tree stems. We do not lose the symmetry even though it of course remains a highly unstable situation. As it turns out, this example of the periodically attracted and repelled billiard ball is equivalent mathematically to the T-tube model, which is equally intuitive but harder to explain.

Is the number of degrees of freedom increased because these obstacles on the billiard table are moving?

We just have a periodic forcing because the trees are all doing the same thing.

And do all have the same mean velocity?

The trees, yes, but the ball, no. The trees are all doing the same thing periodically, and we can assume that they all just oscillate in their width. Then there is this fast particle, the ball, passing through in between. What is going to happen to the rolling ball in the middle? One might take guesses at this moment. It is such a simple game. I should also add that instead of repulsion as Sinai had it, we could have attraction, and ask the same question. But let us first look at the repulsive case a little longer. All these trees are repulsive and we can assume they have a Lennard-Jones type repulsion. Hereby we assume that the attractive short-distance part of the Lennard-Jones potential is omitted. We then have a smoothly repulsive potential left, and its precise form—whether of the 12th order as in the Lennard-Jones potential or whether a repulsive Newtonian potential—makes no qualitative difference. It does not matter whether it is Newtonian, that is maximally long-range, or short-range as in the dip-free Lennard-Jones analog. But we can predict what happens since the situation is so simple. We just have trees oscillating in width on each side, a particle pressing on in-between them, and this repulsive force between the trees and the particle. What we find is that the particle under these forces will be accelerated on average. So the degree of freedom that we are looking at—speed—can be understood in this structurally unstable but very simple scenario, which happens to be identical to the T-tube model. So there is a good motivation to take it seriously.

Can one consider this particle to be a photon?

Presently we have classical-mechanical axioms, so we do not have any problem with the speed of light. But in principle we can include relativity without any problems.

So you believe this picture is equivalent to the T-tube and it can explain cryodynamics. If you now change the model and make it a little more difficult; do these trees or in the billiard table the holes move with the same speed?

All of them move together with the same speed.

If we consider every hole or every tree as a particle, then we have 2 or 3 particles. Now, is it a two-dimensional or three-dimensional system?

We have two dimensions at the time being.

And if we do not have a perfect symmetry and if the symmetry is reduced, are the results then changed? Do we have here a special behavior?

If we find something interesting here in this very symmetric situation, then this will survive if you make the system more realistic because we arranged the symmetry only to avoid unnecessary complications. Whenever we perturb the symmetry, we get the same qualitative behavior. So this is a prototype model, just as Sinai's model was a prototype itself. The model explains a very big range of problems even though it is singularly simple and formally speaking structurally unstable. When we are interested in the question, "What happens to the speed of a particle when it is encountering heavy particles which are approaching or receding?", we can use this model. The symmetry does not detract from the generality of the result which matters. We can show that this particle as it moves through this alley will be accelerated and in the mirror inverted attractive case, it will be braked. These results carry over to the perturbed asymmetric or irregular versions of the billiard game.

Do you believe that the Sinai paper should be read very carefully before you start to smoothen the disks?

Yes, the plane tree problem is a subcase to Sinai's problem in which we introduced smooth potentials rather than the delta function potentials that he used. We therefore get an added observable which is a change of speed.

To get more progress in this way, I mean through the use of mathematics, one should deal first with the original Sinai paper?

This is nothing but a footnote to Sinai's paper, yes.

It is not easy to deal with that long paper?

I mean he took great pains to prove what his intuition showed immediately. So in a sense the paper is very short and simple despite its complexity. Only if you look at the details as footnotes as it were, they are very rich. You can again make many "footnotes" in the plane tree case but the problem itself is very simple. The question is; can we understand or prove that if the trees are doing this, are moving back and forth from the middle of the path or just breathing, that there is a systematic effect that is bound to occur? The answer is: the particle will be sped-up in the breathing alley.

Another question; some people look at statistical mechanics not from the deterministic Boltzmann viewpoint. They start out from many particles looked at in a statistical fashion. They consider two levels of energy, E-minus and E-plus, say. The statistical approach just counts the number of particles lying between two levels of energy. So they do not talk about entropy. We have just two levels of energy, we have many particles, and we are able to count the number of particles. This is statistical physics seen from another viewpoint. What is the role of cryodynamics here?

You see, statistical physics in thermodynamics is just a tool. If you can introduce a deterministic basis underneath the statistical physics which explains everything, yielding statistically valid results from a deterministic basis, then this is a more powerful theory. One can then reduce the number of statistical axioms to zero because you now have only deterministic laws. Chaos theory is a big improvement on statistical mechanics, which fact has not been appreciated sufficiently so far. Here Sinai was the first to show and prove that you can understand a gas in terms of chaos theory, that is, in terms of a dual to celestial mechanics. You do not need any statistical assumptions. Statistics can be used later, once you have understood the deterministic principle. You then can even arrive at proofs why the axioms you introduced in the statistical approach are reasonable.

Chaos yields a lower-level theory, so statistical mechanics is an implication of chaos theory here. And the same applies if you turn everything around and use attractive potentials, which is the sister case to the one which we just had. We had the trees that are moving and could predict that the freshly thrown-in, frictionlessly rolling particle gains energy when the trees are doing what we assumed. This could be shown in detail. If we now invert the whole thing, then the symmetrically thrown-in particle predictably is going to lose energy when it goes through the mirror-inverted alley that is now attractive rather than repulsive. Thereby we come to realizing that there exist two deterministic theories, two many-particle deterministic theories, one chaotic with repulsive potentials, one chaotic with attractive potentials. Both have the same level of importance. It is a strange fact that there exist here two mirror-symmetric sciences, only one of which has been looked at in the history of physics. It is unprecedented that a major discipline of the same level of complexity as another one got overlooked for some 150 years.

Let us finish on a more aggressively formulated applied question if you apply your idea to cosmology: is the idea of cryodynamics at variance with the Big Bang model? Or do you believe in another explanation?

You are alluding to the Hubble law which is an empirical fact. When photons travel through the cosmos, they suffer a distance proportional redshift—a loss of energy which was discovered on a narrow empirical basis by Edwin Hubble and was first hypothetically explained on a deterministic basis by Fritz Zwicky the same year in which Hubble had published it, in 1929. Hubble was during his whole life quite attracted to this idea of Zwicky's—a fact which might have cost him the deserved Nobel Prize. Zwicky unfortunately made a mistake, an error of sign, in his paper, so that the world was not accepting his correct intuition. It sounds incredible, but Zwicky proposed a distance proportional reduction in the energy of the photons passing through a cosmos containing randomly moving heavy galaxies—just like our inverted plane trees. I should say something much more defensive here. At the present moment in time it is not yet clear whether this new statistical mechanics, this new subfield of statistical mechanics called cryodynamics, is also quantitatively able to explain the Hubble law. All that is for sure so far is that it is qualitatively able to explain the Hubble law. There is a paper my coworkers and I published in 2007 that shows that it also fits quantitatively but that is a fairly rough estimate. It no doubt will be possible to arrive at much harder criteria. But at the present moment, one can already be sure, I dare say, that qualitatively speaking, the Big Bang has been superseded. Whether it retains a quantitative niche after being no longer needed qualitatively, is another matter. From a more general point of view, the probability that you have two different principles to explain the same phenomenon—one first-principles, the other based on "second-principles"—is not very high. There are much too many sub-postulates in one of the two.

So, you believe this idea, the inverted alley problem, can explain the Hubble law completely?

I am absolutely convinced, but that is just a bet, not a proof.

You believe that this is a new explanation in place of the Big Bang theory? Is there a point where your viewpoint contradicts the Big Bang, or is it just an offered alternative explanation? What is the advantage of this theory over the Big Bang model?

Cryodynamics makes the Big Bang theory qualitatively obsolete, but only so if this can be confirmed also quantitatively. So there is a quantitative problem that is open and very important to solve. At the moment everything speaks in favor of the hypothesis that Zwicky's explanation of the Hubble law was correct notwithstanding the fact that he at one point committed an error in his equations. If it is true that we here have a new application of statistical mechanics which predicts a Hubble like law, then the first thing we have to check is whether this prediction suffices to quantitatively explain what we observe. Then—if this is the case—the whole of cosmology kind of restarts—is reset to the year 1929 when Hubble had found the empirical law which later proved to be correct. The explanation seen, but not correctly explained, by Zwicky, will then become the correct explanation. So, just as statistical thermodynamics finds a deterministic basis in chaos theory, so statistical cryodynamics has a deterministic basis in chaos theory. The Hubble law is a potential straightforward application. All the beliefs of the scientific community in expansion-related explanations of other cosmological laws like the so-called Cosmic Background Radiation, or like "accelerated expansion," almost everything in cosmology would be in for an overhaul. It is as if all cars that were designed in a realm of physics since 1929 had to be called back to the factory because an element of their engines needs replacement.

Chapter 22
Planck's Constant, Pauli Cell, Indistinguishability, Spin

The concept of spin is important in quantum mechanics. Do you think quantum mechanics can present a complete explanation for it? Does this concept have a correspondence in classical physics as well? How do you explain it? In quantum mechanics, it is somehow explained with Planck's constant. What is your explanation of the spin?

This is a very challenging question, and I think it has never been answered. One accepts the spin as an invention of Pauli's. It is an ingenious discovery. It had no precedent and it needs an explanation. Quantum mechanics as a whole needs an explanation, the lower-case h is also unexplained so far. The spin is in the same category as h, it is a new miracle of nature that had to be accepted and could not be explained. Of course, science always tries to explain what it finds. So the scandal of h, the scandal of spin and the scandal of c, the speed of light being a universal constant, are three scandals of the 20th century which have the same magnitude as one could say. Pauli was responsible for the spin idea. One can have the suspicion that there is a connection between spin and h. To explain h causally is a task that science has the duty to try even if it fails along the way. And h can be explained, but so at a cost. The cost is that nature in some of its aspects becomes private. The individual observer plays a role that is even bigger than one is accustomed to from classical mechanics. We are already familiar with such a phenomenon from c, from special relativity. Einstein was the first to fully appreciate this. Achilles passing by Einstein while a light flash is going off at the latter's feet would have the same spherical cloud of light, expanding around himself at the same speed c, in his own inertial system. Einstein saw the threat that these two private spheres might mean the end of an objective physics, but on the next morning could combine the two spheres as both fitting into the same higher-dimensional structure which in a cross section looks like a cone—the light cone. The latter was in principle known, but its vital necessity to restore rationalism had not been glimpsed before. So there must now exist a similar way to explain h, and with h also spin. There is a classical notion that is not well known which is of great help in this context. It is the concept of "indistinguishability." Everybody believes it is a quantum feature, but this is not the case. Elementary particles are absolutely

A. Sanayei and O. E. Rössler, *Chaotic Harmony*,
Emergence, Complexity and Computation 11, DOI: 10.1007/978-3-319-06781-0_22,

indistinguishable if they belong to the same class. Electrons come in two classes, alpha and beta, but apart from this distinction according to spin, electrons are absolutely equal all over the universe as far as we know, and there is much evidence in favor of this statement. There are very distant clouds, hydrogen clouds, in the universe which have exactly the same Lyman-alpha lines in their spectra as down here on earth. So we know very well that the whole universe is obeying the same laws of quantum physics. It is amazing how homogenous the universe is in this respect. This amounts to a great encouragement to try and completely understand physics—including *h*, including *c*, and including spin. But for some reason, the younger generations over a century gave little or no priority to solving the burning problem of how to explain these three phenomena—*h*, *c* and spin. I have a proposal to make here—I shall try to put it into words. The decisive ingredient is the notion of indistinguishability. It can be introduced as an axiom into classical physics. What is the consequence if you assume to have in a classical formalism absolutely indistinguishable particles? This is not a very new idea, actually. It goes back to Spinoza and Leibniz and has an even longer tradition. Hermann Weyl discovered that Leibniz had already dealt with this question and spoke of the "Leibniz-Pauli principle."

Can you tell in which book or text Hermann Weyl talked about this?

Weyl wrote a beautiful book in 1928 on the Philosophy of Science and Mathematics which has a second edition in English in 1949.

"Philosophy of Mathematics and Natural Science."

Thank you. All three notions are in the title. Hermann Weyl was quite an interesting personality. He found out something very important here as I learned from Hans Primas in Zurich. Primas is a pupil of Wolfgang Pauli's, so there is a direct tradition. I am very obliged to him for having given me this information about the prehistory of the notion of indistinguishability. If we put indistinguishability as an axiom into the classical theory of dynamical systems, what is the consequence? Consider classical phase space—for example, the configuration space with the positions of all particles plotted against each other…

And add time to the positions…

… such that there are not just points but lines since each point moves in time. Time is here not a dimension, only a parameter by which you label the points in such a space. "Configuration space" is if you give each particle a dimension of its own, as you can do visually so long as you have only two particles in one dimension. What happens if you assume that two particles in a frictionless tube are absolutely equal in every respect? This absolute equality is a fact in physics as we saw. We saw that electrons in very distant galaxies have just the same properties as electrons on earth. How come there is this absolute equality? People nowadays talk of the "Higgs boson," for example, and assume that there is a field and this field manifests itself in these particles. But field particles—we are familiar with photons—usually come in very many different varieties, in many different energies, so fields do not provide the necessary explanation for the perfect equality…

So you add the axiom of indistinguishability to classical physics.

That is my proposal.

And after that?

After doing that you find a new classical physics which is still classical but which has new features in configuration space. This is the space we are already familiar with that gives the correct description of the positions of several particles—say two particles on a line. We take our two particles on a line and assume that they are completely indistinguishable. What is the consequence? You still can describe where each of them is located along its own axis at a given moment in time. Take the vertical axis for particle 1 and the horizontal axis for particle 2. Then you can describe exactly how they are moving. But now, add the assumption that they are indistinguishable. That is, you do not know which one is number 1 or 2. It turns out that you then need two descriptions if you are conscientious. That is, you get two trajectories now, two moving points, in the 2-dimensional configuration space spanned by the vertical and the horizontal axis describing the two simultaneous positions at every moment. Every momentary position of either particle on the original one-dimensional line (or tube) is thereby covered. You find that, when the two particles meet (their positions on the original line are equal), the two trajectories—the two drawn lines in your plane—meet on the diagonal. So instead of letting each line continue into the other triangular half-plane, you can assume that the trajectory instead was "mirrored-back" on the diagonal into its own half-plane. This description turns out to be the correct description since the two half-planes are mutually identical. So strong is the power of absolute indistinguishability. It does not matter which triangle you remove because they are identical. So in configuration space, you suddenly have a reduction of its volume by a factor of two when you have 2 indistinguishable particles on a line. This is still pure classical mechanics. If you have 3 indistinguishable particles, it is $3! = 6$, and with N it is N-factorial. So the number by which you must divide configuration space (as Boltzmann called the space of all positions plotted against each other) rises up very fast. Thus if you want to describe a gas of N particles on a line, its ordinary configuration space is an N-cube (analogous to the square for $N = 2$). If you now again introduce indistinguishability, you find that the N-cube is divided, not into 2 identical cells as was the case with 2, but into N-factorial ($N!$) identical cells. This is physically correct and in principle well known since Gibbs' book "Elementary Principles in Statistical Mechanics" of 1902. It means that the volume of your configuration space is maximally shrunken, since N-factorial means a very large number—all numbers up to N being multiplied with each other, a number that rises incredibly fast with N. If you, for example, have a gas of 10^{23} indistinguishable hydrogen atoms in a volume of about one liter, and look at configuration space, then a single one of these $N!$ cells describes the total physical reality. It then turns out that these resulting cells have the approximate size of h per particle (if you also include the velocities in the picture in order to get actions). The equation which describes the whole thing is called the Sackur-Tetrode equation. It gives you the phase space volume in the form of a number which is an action raised to the Nth power. This action represents the phase space volume of the surviving cell.

Notice the nice *finale* of this whole grandfatherly story from the 19th century. The action—call it h^*—is more or less close to h in its value under realistic conditions. So we have here, in an old classical domain the prediction of a number specifying a system-specific action which is close to the h that we are familiar with from quantum mechanics. For example, you can take the volume of your own brain, pretend that it was a classical system with electrons being the main particles—the others for being heavier contribute less to the cell size. Then you will come up with a number for a "unit action" for this classical system which is about $h/20$, which in turn is close to "$\hbar$" ($= h/2\pi$). This classical action is within a factor of three equal to $\hbar$, the main constant of the quantum world. Thus one has here a classical theory that comes up with a unit action that is specific to a particular system but at the same time is very close to the effective Planck's constant. This virtual coincidence can be taken at face value and be proposed as a hypothetical explanation of Planck's constant. This is the only approach—to my knowledge—which tries to explain h so far, and it is quite trivial. Why is this embarrassing? The drawback is that if you change the temperature of the brain, or its volume or anything, then the pertinent unit action h^* which you get is dependent on what you did to this system, the brain, in terms of volume and temperature. So h^* is not a universal constant of nature here, it is only a constant that is specific to a given, momentarily closed physical system. It just happens to be close to $\hbar$ in the case of the brain. In spite of this coincidental character, it is at first sight very tempting to guess that this h^* yields a first tentative explanation of $\hbar$. If one entertains such a hopeful thought, however, one problem arises; how come that the h in physics is universal while our h^* is not a universal constant but a constant only for a specific system? At this point, one can either start doing the necessary calculations more accurately—which is a mess to do—or one can start thinking endophysically. Hereby, one asks the specific question of how a subsystem of a classical universe sees the rest of that universe if this subsystem is on the one hand endowed with an inescapable unit action and on the other is endowed with consciousness. This embarrassing question arises if you believe that the subsystem in question—a brain—is worthy to be honored by the strange question of how "the rest of physics" looks relative to "it" so as if it were of paramount importance.

You want to explain spin. But before coming to this; why do you believe that this unit action h^* can be considered as Planck's quantum of action, since everything you mentioned so far was based on classical physics. We just have a parameter here which you call h or whatever.

I call it h^* written as a lowercase letter as a system-specific action.

We have a parameter here which has the dimension of an action?

It is not a parameter, it is an observable, it is a mathematical implication of a given system.

Why is it Planck's constant?

It is not Planck's constant, it is only a unit action and actions are usually being measured in multiples of h which is the smallest action we can observe in nature. We know of only one constant action so far, Planck's constant—h or $\hbar$. This newly derived action is not h, it can be named h^* (or $\hbar^*$ if you wish). It is a system-specific unit action which just happens to be close in value and meaning to h. So we have the right to give it a name which is related to h.

Now, how can you explain spin?

Now we go over to spin.

You mentioned the Sackur-Tetrode formula.

The Sackur-Tetrode formula explains, not Planck's constant h or $\hbar$, but h^*, a new system-specific unit action that exists everywhere in nature and is different for every momentarily closed system. Let me give it a try. We have these cells in configuration space. They were obtained as a direct consequence of the existence of indistinguishability, of classical indistinguishability. Indistinguishability here means not only that you get a phase space cell called h^* which is different for every closed system—its "Sackur cell"—, but it in addition shows that it is worth having a closer look again at what it means when you have indistinguishable particles in a one-dimensional box. I say one-dimensional because no one ever completely succeeded in doing this in more than one dimension classically, although the problem is a well-posed one. I always tried to find a stronger and younger person to do the mathematical work to solve this open 2-D or 3-D problem. In one space dimension it is easy as we saw because you get these "phase space cells" explicitly that are all identical and of which only one remains. We saw that the size of the cells in configuration space became very small as N grew large.

Let me return to one-dimension with 2 particles again. We assume for simplicity that the two particles do not see each other because they can pass right through each other like solitons. We would not need to make this assumption in 2 or 3 dimensions, but in 1 dimension it facilitates understanding. What we get is two subcells as we saw, two triangles, one located below the main diagonal of our square-shaped configuration space, one above it. Only one of the 2 is the surviving configuration space—it does not matter which one you choose since both are identical as we saw. This half-square yields a complete description of what happens. We now can ask what it means in physics if we return from this reduced configuration space to the real space of nature?

In physical space, we can deduce from what we have in configuration space; that something absurd happens. Whenever the two particles cross each other in real space (as inter-penetrating solitons), the two exchange their identities! This is very hard to believe but easy to show mathematically. So the two particles do no longer pass through each other as we know they can do. We saw that they could be solitons of exactly the same energy being exactly equal mathematically that pass right through each other. What we find is that when the two solitons meet and cross, this is *not* what occurs in physical reality. Rather, in reality each now returns to its own side—into its own "cell." That is, you suddenly no longer have just "phase space cells" (of which all vanished except for one), you also have real-space cells. So in real space, particles suddenly occupy niches even though the whole motion is unchanged. The particles are confined to cells in real space. This is a very surprising fact which was first seen in chemical reality in Tübingen by Walther Kossel in 1916, by the way. These real-space cells can be mathematically proven to exist as soon as you assume indistinguishability, classical indistinguishability. We then get both phase space cells and real-space cells.

Next, you can look at the properties of these unexpected real-space cells in physics. They are very small cells. In a joint mathematical paper with Peter Weibel, a friend of the late Kurt Gödel, we looked at three indistinguishable particles on a ring. On the one-dimensional ring, you of course still have the same local phenomenon that two interpenetrating particles return because they exchanged their identities during the act. But something additional occurs on the ring: two particles exchange their identities also without any bodily contact whenever they are crossing a diagonal running through the middle of the ring. Therefore, we now suddenly also have "jumping particles"—jumping identities of particles. One particle here and one over there exchange their identities in classical physical reality—a most unexpected finding. Imagine, this one here continues up over there and the one formerly over there continues right here. Thus you get this ring-shaped system of "cells" that consists of 2 spatially separated subcells lying exactly opposite to each other. The whole thing, even though classical and exact is fascinatingly magical. It is a wonderful phenomenon. I do not know why it is not very well known: It could already have been seen by Gibbs in 1902, being implicit in what he wrote. But this one-dimensional case stands alone in mathematics so far. The problem has not yet been formerly solved for two dimensions—even for a narrow strip with two solitons—, or in three. It has been an open mathematical question for 23 years and it would be wonderful if anyone was interested in solving it rigorously (there is no doubt that the cell formation continues). We now see that there exists a radically new fact in classical physics—jumping identities of particles—which has nothing to do with quantum mechanics in the sense of being caused by it. On the contrary, quantum mechanics inherits this fact from classical physics. Nevertheless, when we use the right classical cell size, namely the one that happens to be mine at this moment in this brain, then we get even more quantum mechanical features, including h^*. That is, quantum mechanics can apparently be reduced to this classical phase space partition caused by the classical indistinguishability of particles.

Now spin—as you demanded. We already saw that particles live in cells, but these classical cells contain as we saw only one particle each. While in physics, it is almost always two. The reason is "spin." Spin refers to the strange property that in quantum mechanics, we find that the particles live in their cells in pairs in general. This is the difference to the classical case that we discussed; that in quantum mechanics, you have two occupants of a cell in general instead of one. How come that quantum mechanics on the one hand obeys the laws of classical mechanics here—the described cell formation in classical mechanics—but then on the other has two occupants rather than one in general? Spin is the answer. The new property of spin, discovered by Pauli as an explanation of Kossel's finding means that you have electrons of two kinds, "spin up" and "spin down" as one usually says. Every kind of particle if it can have two different spins can live pairwise in the same cell in phase space according to quantum mechanics. Now that we have an explanation of a phase space cell in classical mechanics, we "only" need to explain why it is that there can be two spins and hence two occupants in one cell.

This is the Pauli principle.

This is the Pauli exclusion principle in a classical phase space that we are seeking to understand.

What is the meaning of spin?

In classical physics, we do not have spin. We only have particles living in cells. All particles live in a cell in chemistry and so in physics in accordance with classical mechanics without spin.

In your theory, you added the axiom of indistinguishability. This classical theory does not know of spin?

Yes, it so far only gives you the phase space cells, but it does not give you two occupants, only a single one.

In every cell two particles can simultaneously live?

Classically, you only have one occupant for every cell but in quantum mechanics, you mostly have two.

You now want to explain the Pauli principle in this picture?

Yes, we by now have explained half of the Pauli principle. We explained the existence of "Pauli cells" so far, but we have yet to explain why there are two occupants in a realistic Pauli cell and not just one. This implication arises naturally, too, but to see this, we need to introduce a further step. While for a while we could relax since the disquieting fact of observer-centeredness or endophysics did not raise its head in the above classical description, it here encroaches again. That is, it becomes necessary to include the observer in the picture of physics once more. Physics is then no longer a theory which describes a system from the outside perspective—as indistinguishability theory still does, it becomes one that describes the universe from the point of view of a partial system, Boscovich-style. A kind of "difference principle" enters. I call this "Endophysics"—physics from within as we saw. In this complete picture of physics, you have to acknowledge that you have an observer who is composed of classical quantum cells herself or himself.

This observer is observing the rest of the universe, as we all do. Then you find out to your surprise that this classical observer is subject to yet another indistinguishability; this time it is an indistinguishability in time. The observer cannot distinguish between moving forwards in time at the moment, or backwards in time. We know of course that there actually is no motion in time, that we only believe that there is motion across time, but that this is an illusion from the point of view of physics. There is nothing in physics which says that time is moving at all (not to say forward). This is just something we know from experience, and our models must not contradict this empirical fact but they cannot explain it or even touch it. Subjectively speaking, we are of course moving in time in this inexorable fashion that "is in all" as poet Rilke says. Why we are put onto a new square in time at every moment, we do not know. But we can know that when the particles in our body, or in our brain—where consciousness resides—are reversing the direction of their motion, then these two versions of the observer before and after the change of direction are again indistinguishable. Thus there exists a second much more restricted "indistinguishability" in physics which applies to observers regarding their own direction in time. Indeed the direction in time within an observer is changing in a jump-like fashion from one moment in time to the next, as the particles inside the Sackur cell of the observer are shuttling back and forth. This fact gives you temporal slices. These oscillating temporal slices present inside the observer have the consequence that if you (the observer) look at the outside world, all the particles in the Pauli cells of the world shuttle back and forth without your being able to tell what is their forward and backward direction in time. But why then are there two particles in every Pauli cell—or in most Pauli cells—and not just one? I only once concocted an explanation in an exchange with Michael Conrad. Namely, how does the world of particle motions look if you are subject to time reversals yourself? It turns out that, if you cannot distinguish between your own going back and forward in time, then two equal particles, one moving towards you and the other moving away from you, are no longer equal because of having an opposite direction of motion—spin—relative to you. The correspondingly different angular momenta the two particles have in relation to the observer cannot be distinguished by the latter. Forwards and backwards are indistinguishable, and so are a positive and a negative spin. This would be a deterministic endophysical explanation of spin.

Maybe you can indeed explain the nature of spin from the endophysics point of view, but the point is that in classical physics, can we not explain the quantum spin?

Yes and no. Endophysics is classical physics, with the restriction of the latter being looked at from the inside.

Endophysics is just a viewpoint to explain phenomena?

No, endophysics is a mathematical implication of classical physics. As such it is a reality.

But you said that in your classical model you do not have spin. How can you explain spin after you assumed that you do not have spin?

As soon as you have an internal observer inside the classical model world, one who is subject to these indistinguishable backward and forward rocking motions in time as in rock music, then to this internal observer the rest of the model world appears to have spin. Spin then is not something that exists in the universe. Spin is something which the universe objectively displays relative to an internal observer who himself is moving back and forth in time internally. So the spin would only be an observable within the universe—an observable which pays tribute to the fact that as an internal observer, I am observing this phenomenon. So the spin with two occupants in the same Pauli cell is not something that exists in physics objectively, only something that exists for an internal observer within her or his physics. An endophysical phenomenon, that is.

Can you make a bridge between this idea and the Sackur-Tetrode equation, which you mentioned? I mean, how can the two concepts of spin and h^* be put into a relationship?

Yes, spin just like h is an intra-observer feature infecting the measurable outside world according to endophysics. While h^* was a direct implication of the correct description of the world with the objectively existing Sackur-Tetrode cells in phase space, spin would be a consequence of the Sackur-Tetrode equation being valid for a particular observer's brain being an internal part of that universe. So, provided that the observing brain is for some reason inexplicably distinguished by consciousness and a Now attached to it as two nonphysical features. Then, the rest of the universe would be "distorted" in its appearance toward that observer. Namely, by the "micro-time reversals" that take place inside that observer. Every observed particle then suddenly has spin. Therefore, two particles are now no longer indistinguishable if they objectively move in opposite directions since, to the observer, two phases of her own micromotion in time are indistinguishable. It is a beautiful brain teaser and it actually is still not totally understood mathematically except for one-dimensional observers. It is similar here as with the phase space cells when the underlying space has more than one dimension. One already sees what is bound to be true, but the details remain to be fully elucidated mathematically. So I can put it only in a roundabout way as a conjecture that my own micro motion in time makes it suddenly possible to distinguish between two particles which in reality are indistinguishable. I have not been thinking about this for 23 years, this is the first time that I come back to it. I convinced myself at the time that for two particles differing in their motion in space relative to the observer, you get a differentiation such that the endo observer can distinguish between right-turning and left-turning electrons and that they therefore do both fit into the same observed phase space cell.

Then you use quantum mechanics again?

On the contrary, quantum mechanics is thereby explained with one feature, spin. I just realized that this feature entails an additional prediction. I have tentatively explained quantum mechanics with the little h^* before. Now, these time-reversals in the observer make sure that when I am observing something, I get a "new distinguishability" between right- and left-turning particles and therefore I can now have two occupants in a phase space cell outside of myself. All of this is an objective reality only for me; in the exo reality, there is no spin. Only if an internal observer of the universe observes a member of a class of indistinguishable particles in its classical Pauli cell, then where there should only be one, there are two. It is an endogenous bias present inside my brain that is projected out into the physical world, just as h^* itself is a projection out of my body, and presumably c^* itself is also a projection outwards. So, all these three things—h, c, and spin—come from me, from my being a part of the objective universe. This is mathematically allowed, but physically it is an open question. It reflects this "grip" of being attached and confined to one particular part and parcel of the universe at one particular pre-assigned moment in time—I apologize for the theological connotations.

I do not understand this particular viewpoint of yours. Do you believe in your own thinking that quantum mechanics is a 'part' of classical physics?!

Absolutely, yes.

But this does not make sense to me. In a limiting case, you can offer an explanation, in that limiting case you can have classical physics. For example, if this h^* is going to be very small, going all the way to zero, then we can have a classical viewpoint.

Yes, you then have a classical viewpoint not only from the exo perspective, but also from the endo perspective—at an exact zero temperature of the observer, zero on the exo level.

But this means that classical physics is a part of quantum mechanics, not that quantum mechanics is a part of classical physics?

It means that classical physics is a limit to quantum mechanics, not a part of it.

"A limit" means "a subset."

The limit that you refer to here as a limiting case is not accessible in physics because in physics we have this empirical quantization. But you can also want to explain this physical quantization itself.

But you are saying that quantum mechanics is a "limit" of classical physics?

No, I dare say that quantum mechanics is an "implication" of classical physics. Therefore it cannot be a limit. I mean that from the point of view of quantum mechanics, you can say that classical physics is a limit to quantum mechanics. This is correct. But if you start out from quantum mechanics, you do not understand this fact intuitively; you only know that it is true. From a quantum perspective, you do find that classical physics is a limiting case, right. But you can also look at classical physics as the deeper theory…

…but then you should add some axioms like the indistinguishability axiom.

No, this added axiom in classical physics is valid no matter whether quantum mechanics exists or not. It is an objective part of physics, a classical reality that cannot be doubted.

But you said that you introduce this axiom mathematically?

Yes, but this is unavoidable if I want to describe physics mathematically. I also take the axiom that a continuous line exists, but we need both of these axioms of classical physics to describe classical physics itself. This is absolutely noncontroversial and independent of quantum mechanics. Classical physics from the beginning includes the feature of absolute indistinguishability, even if it was not recognized before except by Leibniz and Gibbs (and Primas) as being a part of it. Only then it turns out that classical physics, with the axiom of classical indistinguishability included, can explain quantum mechanics as an endo implication. This is a different way of looking at it than adding an axiom to quantum mechanics. You certainly can also start out with quantum mechanics, and classical physics then is a limit under certain artificial conditions. Both views do not in any way contradict each other, it is a matter of taste as it were. I only would say that if you can explain something completely in one way, then you no longer have a dichotomy between two equal-rights approaches. For in the one case the price to pay is much higher than in the other. Yet this is a matter of taste. To date, quantum mechanics is routinely considered to be the more fundamental theory, even though indistinguishability (a transfinitely exact feature) is assumed to be a part of it, which is a non-sequitur. On the other hand, many will also not like this privacy of the world of the observer which is brought in by deterministic classical endophysics, which is so embarrassingly close to metaphysical thinking. Physicists do not easily accept having to return to an earlier enlightened era like the pious rationalism of a Descartes or a Boscovich or a Maxwell. The problem is that the observer is given a center place in the universe. This is very unusual nowadays, only Einstein always confessed to be an adherent of this type of thinking which he attributed to Spinoza.

Chapter 23
The Sixth Hilbert Problem, Physics and Beauty

In today's discussion I would like to address some fundamental questions. We have the famous list of the 23 Hilbert problems, some of which have been solved, some incompletely solved, some of which are still open problems, and some have been considered to be very vague. The "sixth Hilbert problem" proposed in Hilbert's 1900 lecture at Paris is generally speaking the proposal to axiomatize those branches of physics in which mathematics is prevalent. The proposal is to treat in the same manner by means of axioms those physical sciences in which already today mathematics plays an important part. In the first rank are the theories of probabilities and of mechanics. We know that in the 1910s, Hilbert and his great assistant Emmy Noether worked on celestial mechanics in this axiomatized fashion. Then, in the 1920s and 1930s, Hilbert, von Neumann and other people as well as Dirac and Jordan worked on the axiomatization of quantum mechanics and the "Hilbert spacei" as is well known. In the 1930s we have probability theory which was formulated by Kolmogorov, and the use of measure theory and other things, tools used for example in modern quantum field theory. The point is that the problem has not yet been solved—I mean it is open. If you go to one of the mentioned cases, for example to quantum mechanics, you encounter different viewpoints. Some of them employ a large number of axioms which does not make much sense. According to Imre Lakatos, such work is very nice, but if you introduce axioms which are artificial and do not come from physics, this does not make very much sense.

It could make sense.

It could make sense if you reduce the axioms and have only palpable axioms, for example, from physics. The question is; do you consider this framework as a consistent physics? Can we have a mathematically consistent physics without any paradoxes or problems arising?

I would subscribe to this idea. It comes from René Descartes who first proposed it. If you wish it is even older, coming from Buddha who claimed that the universe is consistent. Physics is full of surprises. The so-called neutrino oscillations in

A. Sanayei and O. E. Rössler, *Chaotic Harmony*,
Emergence, Complexity and Computation 11, DOI: 10.1007/978-3-319-06781-0_23,

nuclear physics, where particles change their identities over time, no one understands enough to date to come up with a model. It is wonderful that always new riddles are coming up which have yet to be formalized completely. Everything that has not yet been formalized needs to be formalized. Of course when you need to introduce probabilistic theories on the lowest level, this is a big drawback—which one felt has to be accepted in quantum mechanics. There are still attempts to get rid of the accepted indeterministic explanations of quantum mechanics, as endophysics tries to do. The goal of Descartes to find a consistent mathematical description of the universe is still a valid goal and should be upheld in my view. As soon as it is given up, the spirit of science is also given up.

Do you also believe that we should have a consistent physics theory and should follow this avenue?

It is the aim of science to obtain a completely transparent picture of the machine of the universe, as Descartes would have said. This is called "rationalism." Heraclitus, a contemporary of Buddha's, had the same idea. Like these two, Descartes had a very humanistic motivation for proposing this machine theory of the universe. Its purpose was to rule out that we live in a world that is controlled by potentially evil spirits—which would be the alternative. If you really think there exists a so called "primary chance," as Wolfgang Pauli called it, then this means that witchcraft is built into nature. Witchcraft is of course the opposite of rationalism. So quantum mechanics came very close in history to sacrificing rationalism. It is an interesting fact that this was not taken up as the challenge it presents: to find a way out of this trap as soon as possible.

Quantum mechanics in the canonical Copenhagen view acknowledges that science so far does not and indeed cannot predict the point in time when a certain Geiger counter will click because of a quantum transition occurring on a probabilistic basis in a radioactive probe. These clicks could be influenced by witchcraft in principle, and quantum mechanics would have to subscribe to this witchcraft if it could be identified as a viable force in the world. I once proposed an experiment for public TV called the "Hot Chair." A person—anyone who is ready—is invited to sit down and be confronted with a "Dehmelt atom" that is producing little light flashes irregularly on a quantum basis in accordance with the probabilistic equations that exist. There would be a little computer installed that translates the light pulses of the Dehmelt atom into a Morse code. Each combination of light and dark intervals would give rise to a letter, or most of them. The Morse code would thus give a real-time translation of the activities of this flashing atom that everyone including the audience can see. The task of the person sitting on the hot chair would be to prove to the planet that she or he can influence this Dehmelt atom, the signals that come out from it. This would be an absolutely acceptable scientific situation. If a person hereby proved to be able to make a meaningful short message appear within the otherwise meaningless salad of letters, of say ten or twenty letters in length, then this person would have to be officially made a part of the laws of nature. So this example—for which probably no TV station would give money at present but which is in accord with quantum physics as we know it—demonstrates to the eye how much the progress of science has been imperiled by the probabilistic interpretation of quantum mechanics. Such, that physics itself is no longer immune to becoming part of a belief in witchcraft. It is not known to the public at large to date that probabilistic quantum mechanics stands in violation of rationalism. Because of this accepted silence, rationalism has been abandoned in 1927 with the advent of quantum mechanics—unless one takes seriously the attempts to explain quantum mechanics deterministically.

Do you think that quantum mechanics is not mathematically consistent? What is the problem?

It is at best marginally consistent because it is based on acausal probability as illustrated by the Dehmelt atom.

Do you—to go one step further—believe that we can have two given theories in two different fields of physics that contradict each other in some points and remain so?

Either the one or the other or both would need to be corrected.

Do we have this situation at present, or is this not the case?

At the time being, quantum mechanics and general relativity cannot be unified so far.

But they are not put in doubt, neither of them. One does not see any direct and sharp contradiction at the moment.

Oh yes, I would say they contradict each other massively. Otherwise they could be unified. The fact that they cannot be unified so far means that there are manifest contradictions at work. So physics is at the present moment in time not in a state which can be…

Maybe they can be unified, and we only have not yet found out how to do it. This fact does not mean that the two theories contradict each other, or does it?

It is quite normal that one is not finished finding the right explanation for something. This is the case here.

Consider specifically general relativity and quantum mechanics. Should we go along with Hilbert's viewpoint and axiomatize both theories and then link them together?

If a theory is wrong, axiomatization does not help at all, it only makes it all more wrong because everything is even harder to change afterwards. Axiomatization is dangerous if you do it too early.

What do you mean by a theory being "wrong"?

A theory can be right or wrong as far as nature is concerned. If it is beautiful but wrong, it is a great pity that people will be misled more easily into following the wrong road.

But I think we cannot use the words "wrong" and "right" for a physical theory. We can only say that it has not yet been falsified.

Okay, but these are shades of meaning. Popper was of course a very bright and important mind, but this insight of his is a footnote at this point. If you find a theory which works—and this means that it explains also a great many little things that were not even seen before—, then you have made very good progress even though you always have to accept the possibility that a new version of the theory will eventually contain this one as a special case while being much more beautiful itself. So a confirmed result of a scientific theory always survives as a special case to a better theory.

Consider quantum mechanics. Since its start almost a century ago by now, this theory more or less works from the point of view of experimentation.

Yes, it works beautifully within the limits which it describes itself. Its drawback is, that it cannot predict when a certain "quantum decision" will be made by nature, and, that certain superluminal correlations exist as in the Bell-Aspect experiment.

We can start to axiomatize quantum mechanics, as a lot of people have tried to do, some with more, some with less success.

Yes, some attempts are more elegant than others.

On the other hand, do you consider that this theory is somehow acceptable so far? And general relativity is as well?

Quantum mechanics is quite acceptable, I would say. One "only" has to find a causal explanation still. As far as the facts—the effects that it does predict—are concerned, quantum mechanics is the most successful and precise physical theory of history.

And general relativity is also a successful theory so far?

Yes, although it is not a very well-confirmed theory as of yet. There are quite a few things that got confirmed marvellously but do not really touch on the core of the theory. The majority is already implicit in the equivalence principle of special relativity from which general relativity was derived. However, there is a major exception—black holes. They seem to exist. Think only of the 4-million solar-mass enigma at Sagittarius A* in the center of the Milky Way galaxy, with its many beautifully fitting confirmed properties. Black holes have additional properties—for example, that they are never finished in finite outer time—which are so far not appreciated to be implicit in general relativity, since the latter has not yet been "tuned" in the right way so that they can also be derived. So far, these new properties only follow from a "pre-version" of general relativity, the Einstein equivalence principle and the famous Rindler metric, but not yet from general relativity in the canonical form that exists so far.

If we want to find the bridge between quantum mechanics and general relativity in order to make physics more consistent; do you consider the possibility of taking some axioms from quantum mechanics, and then some axioms from general relativity and let mathematics find the bridge between the two? Do you consider this method acceptable, or do you have a different proposal?

This method will be wonderful if it works. But you can never be sure that a brute-force method will work. One almost always first needs a new way of looking at some element of one or the other theory that looks foreign at first sight to enable the unification. Just by hoping that by continuing with things that are already known they will suddenly fit together does not work in general, this is wishful thinking.

Then what is your idea as to how to get a consistent physics?

I mean, physics is never consistent in a finished sense. One can always remain optimistic that it will be possible to progress further in the direction of making the whole field even more consistent. That is the problem befitting science since Descartes' book "Meditations on the First Philosophy" of 1641, published after his young daughter had died, a fact which made him fearless. This is quite a long time of continuous progress. One always has to be very modest in one's claims. On the other hand, one has to be very strong in one's beliefs. This is possible because so much depends on rationalism. Witchcraft is maximally dangerous if it really re-enters into the world, even if its proponents are maximally sympathetic people. I have a friend...

Another question. Sometimes mathematics predicts that some phenomena exist which so far have not yet been seen or confirmed. Take, for example, Dirac's hole theory or his prediction of the positron. Dirac created it as something that only follows from his marvelous equation. A lot of people disagreed with it after its first proposal, but it was eventually largely accepted. He said that because the mathematics showed it to me and the mathematics is beautiful, I believed in it. By the way, I do not have a definition of "beautiful"—it is somehow case dependent. Mathematics convinced him to believe in it for physics. It is the fault of the experimental physicists if they cannot confirm it. Today it is largely accepted and we have other things that "merely" come from equations. Consider for example the equations of gravitational waves.

The last word always belongs to empiricism. A prediction that has not been verified (unless it followed from known facts in a faultless logical way) cannot be accepted as a true fact until experiment has confirmed it. The gravitational waves that you mentioned form a typical case in point. They have never been found. Only indirect evidence exists, and there are alternative ways to explain the same evidence. People do not like these alternatives at present to the point of refusing to acknowledge that a relevant parameter in the description of the observations was put equal to zero and hence does not show up in the descriptive equation (as I witnessed). This of course happened unintentionally—the fact that "tidal friction" and "magnetofriction" were omitted in the model. So the "indirect evidence" that exists so far in favor of gravitational waves is insufficient in my eyes. Einstein himself was never quite convinced of their existence. I hasten to add that I would not have been able to spot these weak points if I had not strong theoretical evidence against their existence—namely my new "global-*c* result" in the equivalence principle and in the Schwarzschild metric of general relativity. I do not insist on them in any way here but only say this is the nice thing with science; that everyone is allowed to place bets about how nature works, bets that are subject to falsifiability. Hereby the more elegant shorter paths are unfairly privileged. Everyone has a chance and it is only beauty—elegance—that allows the winner to reach the box first. Students are usually not told this secret of Dirac's.

If reasonable mathematics implies a phenomenon like gravitational waves, should one, as a physicist believe in it or not?

At first I would tend to say, yes. But there is a trap here. If you interpret facts in one particular direction because you like a theory in a simple naked form, it can happen that you neglect other ways to interpret the same data. The method you just sketched is not at all a surefire way to make progress in science. Let me come back to gravitational waves. Hulse and Taylor won a Nobel Prize for the discovery of a pulsar circling an invisible partner. They interpreted their data in terms of gravitational waves—the fact that the orbit slowly shrinks. But they forgot to say that they had put one parameter equal to zero—tidal friction in the invisible partner star if it is a dark white dwarf. I recently met the discoverer of the first and so far only "double pulsar" with its new bonanza of data, Michael Kramer, a discovery which again deserves a Nobel Prize, but again only for the data, not for the interpretation. Alfred Nobel was very sensitive here himself. The "frictional interpretation" of the Hulse-Taylor result has yet to be ruled out. Kramer's stronger friction effect may analogously come, not from tidal friction but from magnetofriction between the two neutron stars circling each other at the earth-moon distance. If this alternative source of friction can be ruled out, gravitational waves would become much more probable.

You do not believe in these observational physical results? Only if the theory says we should have such a thing you can accept it?

No, no, facts must be accepted. Only theories can be wrong, and if a wrong theory explains it all…

… meaning with reasonable mathematics…

…we have a problem, yes. Even with reasonable mathematics there is no guarantee that the theory is correct because of the potentially large set of equally relevant parameters omitted. This is why the experimental method was originally invented in the first place—to reduce the number of theoretically possible parameters.

What do you mean by a "correct theory"?

One being observed by nature herself.

Take a theory that has not yet been observed in its consequences in nature. Consider again Dirac. He arrived at some equations for some phenomena which led him to accept a "negative energy." No one believed him but he said that he could explain it by creating his "hole theory." Some people thought it was a crazy thing to do because we do not see such a thing and experimenters have not yet confirmed it.

If you have time reversals, as apparently exist on a very short time scale in quantum mechanics, then this would be one predictable implication. I am not at all opposed to it because I have a mental bridge of my own.

It was confirmed finally by some experiments.

Yes.

At that moment in time, would you have said I accept your theory, or not?

I admit I would have had reservations. One can never accept a theory as long as there is no plausible evidence that this theory could or must be obeyed by nature. Dirac was one of the most creative physicists in finding new candidate laws that nature might obey. It was perhaps because he had been trained as an engineer first. But as long as such a candidate law has not found confirmation—it could also be a theoretical, very indirect confirmation or consistency with another theory—, skepticism is allowed. So, the nice phrase that "the benefit of the doubt" must never be withheld remains valid. That is, any way of testing the theory must legitimately be kept open and sought. Oftentimes the canon of solid physical facts contains side features that miraculously fit. Thinking creatively, therefore, is always allowed. You need such minds.

If you have an experiment that confirms a theory, this does not imply that the theory is true. I mean, the word "true" is not okay here. We can say "this theory was successful in this experiment" but we do not know about another experiment.

Yes, but some theory can be very convoluted in its logical structure. If it then is confirmed in many ostensibly unrelated points, this automatically also throws light on other previously unseen and still unconfirmed points. So there can be webs of theory that are virtually incorrigible owing to their multi-tieredness. They still might eventually be generalized further and shown to be special cases to a more general theory, but some features will then survive. The "phlogiston theory" of fire and the "miasma theory" of infection did not survive but their correct elements (oxygen and bacteria) did. Also many features that are to date explained with cosmic expansion may someday survive under another umbrella. So the opinion that theories can always be overthrown is not quite correct. One cannot overthrow facts that have been verified many times.

Do you believe that without a theory, we can still observe nature?

You cannot even use your eyes without using theories. So the idea of a theory-free observation is untenable. Any observation is theory-based, including an observation which betrays the opposite. You cannot escape thinking. Hence one should be quite happy with one's own optimism in believing in theory. Theory-building is a wonderful invention and it has been miraculously successful for a million years. I am thinking of a "modern" hand axe that is a million years old that could not have been built without theoretical knowledge.

A final question; you said that it is not favorable for physicists to formulate the mathematics of some creation of their mind "too early." You seem to believe that as long as the experiment cannot give some positive signals, you cannot believe in the mathematics?

This is not exactly what I would say. There is a difference between a basic insight which can be maintained from the beginning and the crafting of more detailed theories which need time to blossom. So my answer has two wings as it were. Some mathematics is presentable soon and should be worked out before too much phenomenology has been accrued.

Then what do you think of this saying of Eddington's which you mentioned in your course: "I never accept any result of experimental physicists unless they fit my theory." This is the other way around?

It is nice, it was a bit tongue-in-cheek, was it not?

What do you really think about it? Are you sure it was a joke?

Of course.

You do not take it a little bit seriously?

I mean he was a little bit conceited as he shows with this joke. But he also was joking about himself, which is sympathetic.

But this was not a joke—on the part of Dirac. He somehow believed that if the theory is "beautiful," then it should be taken very seriously.

Einstein said a similar thing once. When you have accrued many reasons for belief in something consistent that you or someone else found, you can stick to that against early evidence which seems to contradict it. Evidence never has only one meaning, so evidence as such is not a very strong argument, at least not at the beginning. Evidence tends to be overrated—and underrated, since sometimes the tiniest evidence is infinitely powerful.

This belief in one's own theory is somehow a metaphysical thing, or is it not? Suppose Einstein had once said "because it is beautiful I accept it". Would this part of the sentence—"because it is beautiful"—have an exact meaning?

It would mean in the assumed case that the theory has so many implications—even infinitely many perhaps with a few of them already known to me—, that otherwise this theory would not have congealed in my head. The chances therefore are very high that it will indeed be confirmed in further observations. But it is hard to put oneself into Einstein's mind.

The same applies to Dirac when he said; "Physical laws should have mathematical beauty." One can personally accept this. But if we can continue a bit more here, what is the meaning of beauty? Can beauty be case-dependent?

Beauty is perhaps the wrong word; one of its main features is simplicity. And simplicity is always beautiful by definition.

What is the meaning of simplicity?

It is just as difficult to define as beauty is, but it sounds a bit more scientific. To quote Newton "Truth is ever to be found in simplicity."

There lie a lot of challenges in the word "simplicity" in philosophy. Many definitions of simplicity do exist.

But simplicity is a form of beauty. It is more easy to absorb than the word beauty which has many connotations that have nothing to do with science. On the other hand, of course, if lay people are drawn into physics by the use of the word "beauty" or "charm," this is legitimate, too. So I would not say that you have to prune physical theory from words like "beauty."

Thank you for combining simplicity and beauty.

Chapter 24
Physics and Religion

In this session I would like to ask a somewhat different question. It actually is motivated by Einstein. Can one expect a good influence of religion on the study of physics and its progress? Often one finds the opposite opinion among physicists—that as soon as we have progress in physics we do not need religion anymore. Some go as far as to say that we do not need God anymore for explaining the whole thing. I would like to ask you, what is your opinion in this respect and do you believe there is an interaction between religion and physics? Can they mutually motivate each other or are they completely insulated from one another?

This unexpected question I find important. Most people interested in science have seen the book "Subtle is the Lord" written by Abraham Pais about Einstein, in homage to his famous synonymous saying. Einstein always tried—in the footsteps of Spinoza—to have a "complete picture." Of course when you insist on having a complete picture, you are neither a scientist nor a religious thinker but just a human being. This is the basis from which one has to start out. There is no difference at all between rational thinking in religion and rational thinking elsewhere. Yet a complete picture is what one needs. Only if one is not afraid of the biggest questions can one suddenly find a new avenue within any given field. The upshot is; there never was a contradiction. In the past, the very good thinkers often were specialized both in religion and in science, Maimonides for example and his predecessor in Persia, Avicenna. These people were originally applied people, physicians, and at the same time theologians. The book written by Maimonides, "Advice for the Perplexed," has a very nice title and covers both sides. Bringing the two fields together became, half a millennium later, the big contribution of René Descartes. Descartes was a religious thinker in the first place and laid the basis for modern science which follows him up to this day. He posed the question of whether the world is consistent as a machine, as a way to get clarity about the big dream that we are dreaming with our waking consciousness at every moment, and whether this situation is acceptable. The fact that consciousness is not a part of science shows that everything that is important is not part of science. Hence there is a more rational way to approach the most important things than that of science. There are facts that have nothing to do with science but are much more

A. Sanayei and O. E. Rössler, *Chaotic Harmony*,
Emergence, Complexity and Computation 11, DOI: 10.1007/978-3-319-06781-0_24,

primary and palpable than any scientific fact—like that we right now have this particular clock time of 1.04 p.m. there is no way to explain this, it is "just" a fact. Where does the Now come from? Color as well, and light itself, brightness is nothing that exists in nature. I once talked with Bob Rosen about the red rear lights of a car driving in front of us, this impressive redness when the car is braking. Redness came up as a topic, and he told me about the founder of the famous young quantum physicists' school in Göttingen…

Max Born…

Born wrote, so Bob told me, in one of his books about this problem of color and its non-existence in physics, a fact that not many people know about. Color does not exist in physics, the Now does not exist, and pain and pleasure do not exist in physics. This, even though there are equations for artificial brains that if implemented physically determine a response like having pain or pleasure but nonetheless are only governed by space- and time-dependent potential gradients generated by endogenous and exogenous clues. What is a potential gradient, an endogenous force? It only has a meaning in terms of subjective experience. So the only reality that exists for anyone is the soul, which is a subjective experience. The latter is everything, all the fears that one has, all the hope that one has, all the skepticism one harbors is subjective. None of these forces exists in science as an identifiable recurrent reality, notwithstanding the fact that one is threatened by death at every minute. One somehow does not talk about this or think about it even though one reads the newspapers. One sees on TV reports other people's problems and wars, and about the danger of the atomic bomb and the danger of black holes, perhaps. So everyone knows about the fragility of life, and how peculiar being inside one particular body really is, although most of the living ones have the diagnosis of a deadly disease or the systematic unfolding of a fatal accident still ahead of them. The body has its problems that need to be cured by different types of doctors, starting with the teeth, with a new hip becoming virtually mandatory in old age, and so forth. But this is not usually seen as a part of science in the sense of it needing a deeper understanding. "Understanding" comprises science and religion, physics and subjective experience. So does the theory of love. Love is not part of science, even though all the elements of it form a part and topic of science. Recently Pope Francis said nice things in South America to young people encouraging them to wake up and to do something, to change something, to take responsibility. But responsibility is again something that does not exist in science. The very idea of being responsible, being able to act as a person and thereby work a miracle, as Lévinas showed, is not addressed in science so far. Nevertheless science is a field designed and opened up with the express aim to work "miracles" by using the leverages laid into our hands. Finding a new thing in science is tantamount to being given a chance to work a miracle, compared to what was possible before. Finding new truths as a precondition is the aim.

There is a letter written to Einstein by a child of 10 or 12 years who asked Einstein, how do scientists think about religion? The letter is part of the collection of letters written to Einstein. I believe that if you asked the same question to physicists nowadays, many would say that as soon as we learn sufficiently much more, we will be able to dispense with God. But Einstein responded differently to the questioner. He answered that as we learn more about things, we also improve our thinking about God. Even though we might get rid of some naïve viewpoints regarding God, we improve our understanding of the non-physical. That is, we learn to understand God on a much deeper level.

This is a nice way to say it.

But my question is; why is it that when we learn more things about nature and in particular physics, our belief in God becomes deeper? And why do others say that as we learn more, we need God less?

This shows that there exists something like two schools here—the skeptic school and the open-minded school. The latter of course has its roots in the old texts, like in the old testament, where there are lines which say how much the speaker feels oppressed by the "fist" of the creator being so close to his neck. One is totally at the mercy of a power that one hopes has a friendly face. It is the insight that one is so dependent on every little seemingly chanceful event in the world—as dealt with in chaos theory, for example—, and of one's being enclosed in this particular skin and having a wart somewhere in the face, or a tooth ache or an ailment.

Why does my consciousness apparently reside in certain cells in this particular single brain at this particular single moment in time? Why am I attached to this body at this moment? I call these the "Assignment Conditions" (A.C.). If you try to have a complete picture of the world, as is the task of natural science, then you find you have the Laws, the Initial Conditions (I.C.) and the Assignment Conditions (A.C.). Why is a certain participant in the world machine located at a particular place in it and assigned this particular consciousness to her or his body or brain or parts of the brain or specific brain cells or subelements of such a cell? These elements exist despite the fact that this same brain might suffer a stroke the next minute. The whole thing is such an—one is tempted to say—"unfair situation" that one cannot help but respond morally to it. Maybe the most efficient way to avoid responding for a modern person is to be brain-washed into no longer believing that there exists anything outside the narrow field of the relations described by science. I am referring to the Assignment Conditions which are never mentioned. Following the car accident which took the life of my young son, the idea of assignment conditions (A.C.) was the first new idea my mind formed in the hospital during my bodily convalescence. Then I knew I should go on. He had called me tenderly "stupid old gray one."

Science today is a substitute religion, that is, is in competition with real religion. But it is not a religion; it is just a technique Descartes invented. He invented it to fight, to vanquish a potentially evil demon. He wanted to know whether this whole thing of being a living consciousness he found imposed on him and presumably everyone else is not "a bad dream." It is a dream since its substance is mind-stuff, not thing-stuff. It is a seemingly consistent dream or chain-dream, but it could be a bad dream, a malevolent imposition, and the outflow of an evil intention. Even though everything is so beautiful with its colors, we also have pain, we have wounds, and we have losses. There must be an excuse or a promise of an eventual healing behind the things that are so hard to accept or even unacceptable. There exist things that appear to be unacceptable, and there are situations in life that cannot be tolerated. Hopefully one is spared them. Religion would have the purpose of giving an excuse for the proneness towards, and the sometimes inescapability of, tragedy. Anyone who says I do not believe in tragedy can also say I do not believe in religion. But it would be stupid to not believe in tragedy, right? It is absolutely wonderful that one has this little niche of science in which people have the assigned privilege of being allowed to work miracles for and do good to their fellow human beings by finding something new that is made available to everyone. Art is on the same level, you can also work miracles there, but you need many personal gifts and good teachers to become a good artist. In science, it is the same thing and you need to be fearless. But you also need to meet very good scientists. You need someone who, for example, tells you while you are still at a fairly young age that you are a "genius." It is an old trick used by the Shamans. It is very much motivating for the rest of one's life. This is, of course, just a special case of interpersonal kindness in general; one should cultivate this kindness which is of the essence of science. It is like the Islamic school in Cordoba where people still knew about the deep helpful mission of human beings within a religious context. Their picture which was much more rational than modern science had influenced as we saw via Maimonides and Saint Francis, René Descartes.

Descartes was directly influenced by Ignatius of Loyola, the founder of the Jesuit school of thought because he attended a boarding school run by Jesuit priests not long after Ignatius' passing away, in La Flèche in Southern France. Loyola invented the "experiment" in a modern sense before Galilei. He did not use the term "experiment" which had not been invented at the time, but the related term "exercise." The famous exercises ("exercitia") of the Jesuit order (that had been founded by him as a counter-reformation to Protestantism) would subsequently acquire a more restricted tamer meaning. Originally, "exercitium" just meant doing a decisive experiment. What would be a really important experiment? Ignatius used his own life as an experimental tool. Like his belated pupil Descartes, he was forced to find out whether the whole thing is acceptable. He had experienced the ordeal of his leg having to be artificially broken for several times following a horse-riding accident without anethesia. Was the gift of life something benevolent in the sense of giving him a role worth fulfilling despite the ordeals? First, he made the experiment of becoming a begging monk. The experiment went still farther as an "exercise" even though it sounds ridiculous to modern ears and disturbingly sweet. He had decided to go to Palestine because he had always regretted not being Jewish and thereby closer to Jesus. So he went to a seaport, mounted a ship and asked the captain, “Please take me along since you are leaving for Palestine in a few hours.” The captain would of course ask what he could pay. You see that I am a mendicant without money. Then the captain would ask him, amused, “Okay, why should I then take you along?” The self-confident answer was: “Because then you will arrive.” This was his way of experimenting with his own life. Two captains would throw him off their ship, the third took him along, or so the tale goes. One can bet with history’s hindsight which one of the three ships was the only one that arrived in Palestine. It was a stupid, a crazy example of doing an experiment by throwing one’s whole existence in as a token. But it was a token played, not against reason, but already within the scope of the daring Cartesian program to test the “degree of benevolence” of the instance that screened to him the living dream of waking reality. It was a very causal type of thinking in spite of its seeming irrationality. Descartes sharpened the test to checking whether or not the world is relationally consistent—causally consistent, thereby inventing the program of science. As long as it still is causal, everything is fine because then one can “put the dream-giving instance to shame” by one’s acting fairly towards a fellow person since the others would be “mere machines” that one would be infinitely privileged over by the possession of consciousness. Not to misuse this infinite power of exteriority—staying outside the mere machines—would cause an evil dream-giving instance to self-destruct for having been dwarfed. But so only as long as the hypothesis of the world and everything in it being a causal machine was not falsified. This was the birth of science in the modern sense, and quantum mechanics still is its deepest challenge. Benevolence here is not a nicety alone but is the sharpest sword conceivable—directed not against the poor fellow person whom one helps, but against the dream-giving instance itself. Quantum mechanics is the ship to mount and safely steer back towards rationalism.

We arrived at the word benevolence. You have co-authored a book about this topic. My question is this; what is the meaning of benevolence? And what is its relationship to religion and physics?

Benevolence is an invention that has been made by human beings.

And love is benevolence?

There is a close relationship but they are not identical, benevolence presupposes love. Benevolence is a notion which got largely forgotten much like religion. Everybody still knows what religion is, that it is not science but nonetheless not nonsensical. Benevolence is a fact, an experience that everyone has made in her or his lifetime. For example, one may encounter it in a bus, when one is carelessly riding without a valid ticket which happens from time to time to almost everyone. When one is then caught and a neighbor whom one never met before says, “This is his ticket” producing his own, and this person on being asked to produce his own ticket fumbles and eventually says, “I must have lost it” and goes free because this is too difficult to disantangle for the conductor. This is benevolence—“exteriority” in the sense of Descartes and Levinas, practiced as a weapon against heaven. It does not take much to save someone else from losing his face.

What is the motivation here? The other person sacrifices himself?

By no means, I assumed that the risk is very minor. But why should one be benevolent, this is the question. Benevolence has to do with being moved in the heart.

You believe one cannot learn it by practicing? Is it a natural characteristic of some people?

On the contrary, I would say that it is absolutely learned. There is nothing non-learned in it, but it is a reality if you had good luck yourself. Actually, everybody knows what benevolence is from early childhood on. For example I remember a very big hand looming over my head. It must have been the hand of my father's but I was not afraid of it. Benevolence is a hypothesis that in human beings comes up at a very young age. This event, when it was first recognized or rather invented, is later forgotten in general. It occurs between 1 and 2 years of age usually, and it is the deepest experience that one has ever made. The experience of benevolence felt to exist turns one into a person. So personhood is something that can be scientifically defined. If you understand benevolence, you are a person. Benevolence, even though it looks very religious, is something which is an implication of a scientific theory called "person theory." We already talked about that at length. Person theory is a theory that can be applied, not just to natural brains—human and animal brains—, it can be applied to artificial brains as well. Thus, science touches here on something very deep when looked at from a philosophical point of view. Physics and life come dangerously close.

Is this an epistemological theory? You talked about the phenomenology of persons, and your theory of benevolence makes a connection to philosophy. Maybe, you want to unify everything here?

I have no intention to unify—it happens automatically.

You mean it is the phenomenology that you just explained?

Exactly, it is a reality which one can understand.

A further question; do you personally believe that religion can be a good motivation for doing physics?

Absolutely, I would even go so far as to say that without religious thinking, you cannot do good physics. Of course, except for spotting a mistake made in an equation written by someone else—that is not something deep enough although it is also a very important scientific achievement. Everything new that is real progress and not just happenstance progress will depend on deep thinking. Deep thinking means thinking as deeply as a child 18 months of age does—which is very hard to do when you are grown up.

On the other hand, if you learn or create something new in physics, then this can have a feedback onto your religious viewpoint. What is this feedback? Why can this feedback for some people decrease their belief in religion while for some others after creating something new in physics their belief in religion is strengthened?

It may depend on temperament. But it also depends on whether your mind has been opened up before towards this deeper level of reality—of realizing that color and the Now are gifts. One has the highest level of theological capabilities as a young child. Some people had the good luck of being nourished in this type of thinking during their childhood and adolescence while others did not have this privilege. Also many people lose this feeling because they think it is not rational and therefore admissible when growing up. They of course keep it in their personal lives and in the way they treat their neighbors and in the character they display and so on, but they do not know any more that this is applied religion. For example, many physicians are such somnambulatory heroes. They think that it is just the behavior of a certain self-confident attitude and do not realize that the principles they use in their lives are intrinsically religious.

My next question is this; I think that you agree that ethics is something more than religion. I mean, a person cannot be religious but she or he can nonetheless be completely moral.

Absolutely, I would say this represents the practical side of religion, and the connection to religion can be masked or forgotten.

You can somehow say ethics is a super-religion because you need not believe in some special things or religions in order to have ethics. You can possess ethics, you can be moral without religion, I think.

In the time of Goethe there was a proverb: "Who possesses science and art also possesses religion; who lacks these two may be given religion." It was a very haughty phrase. If you use the word "religion" in the sense of different religions, of partial religions so to speak, one can still feel the enlightenment's verve behind it. But what I referred to was the real religion which comprises all religions and is behind all religions. It looks like a fight with heaven as Descartes and Levinas showed.

What is the role of ethics in science? A scientist should be moral. On the other hand, let us suppose the following situation; one has arrived at a new creation in physics which is completely new and important and technically appealing but represents a result that could prove dangerous for humanity.

A "technically sweet" bomb, for example.

What is the responsibility of that imagined scientist? Should she or he not talk about it, not publish, because it may cause a problem for humanity? What is the role of ethics in physics?

This is a very concrete question. I happen to know about a real-life example here. My late friend Carl-Friedrich von Weizsäcker was asked by Lise Meitner (who happened to be his boss at the Kaiser-Wilhelm Institute as I later learned, but she may already have been in exile in which case this happened on the phone): "Shall I publish my finding that the "mass defect" in nuclear fission (that she and Otto Hahn had jointly discovered) implies a well-defined large amount of kinetic energy by Einstein's mass-energy formula?" von Weizsäcker had answered, as he told me in the presence of his mother when I visited with him in 1964 in her house near Lake Constance, that he had advised her to publish. It was a strange topic to begin a conversation because I had not yet met him in person and had expected we would discuss chemical evolution. The question posed to him was, maybe, the most important question raised in the 20th century.

What did she do?

She did publish. Only it turned out that her paper appeared 1 month after that of her colleague Otto Hahn (his in January, hers in February 1939) who later won a Nobel Prize for it.

You believe that Otto Hahn was not morally pressed as to whether or not he should publish? That only for this lady it was a gnawing question so that it took some time for her to come to a decision? That this is what we call ethics?

She was a religious person as a child as one knows. Nevertheless she decided in the way she had been advised by her 26 years old assistant who had convinced her to publish because it would become known anyhow—a straightforward idea.

But did she know that Otto Hahn would publish the idea 1 month before her?

I believe not. The exchange with her coworker took place months before. So her decision to publish probably had nothing to do with Otto Hahn's decision to publish and to publish without her, which is hard to understand. She would keep him in high esteem and almost forgive him. The question is, of course; was this good advice or not that Carl-Friedrich von Weizsäcker had given his boss? Why did he say she should publish? And why did he tell me?

On the other hand, it can happen in principle that you will come into the situation that you have to ask the same question yourself from a younger person: Shall I publish a result that, as you said, is "technically sweet" but potentially dangerous for humanity, what would you do?

If the dangers are as big as they were in the mentioned case, it would have been wiser not to publish because the "sin" would then not become your own. In later life she was indeed haunted by the fact that she had acquired a co-responsibility for the bomb. I recently saw a TV documentary about her. She was as a child told by her grandmother not to play the piano on the Sabbath because this would be punished from heaven directly. She promptly put this to an experimental check—and found empirically that nothing happened. So she performed a religious experiment as a child. I had not seen the connection to the story of Ignatius before.

Let me ask you whether you think that religion can be a motivation for doing physics? Can results coming from physics contribute to improving a person's belief in religion?

Yes, religion can motivate scientific success and can also gain from better understanding the subtleties of the Lord to use a pious man's words.

If ethics can postpone a technical publication, do you consider this as something completely positive and fitting?

Absolutely, yes.

So you consider a postponement in science to be potentially necessary. If you have an idea that is dangerous and ethics says you should not publish, then you arrive at the conclusion that you do not publish it, right?

Exactly.

Then science is inhibited by the postponement, or is it not?

We do not know beforehand. It could take a 1,000 years until a theory or idea resurfaces in science.

One must be a very courageous personality to decide to do so and ignore an achievement and therefore, perhaps miss a prestigious prize.

Yes, but this automatically follows from scientific ethics anyhow. Actually, this question is not as foreign as it sounds because there are many very dangerous findings in science that are lying on the table and no one is misusing them.

This is what you believe when you define science; you believe that science is friendship.

Yes, its root is friendship.

And this friendship has for its center ethics.

Ethics is part of it, yes. I once had the opportunity to fly in a tiny airplane from Santa Fe to Los Alamos for a chaos meeting of the University of California taking place in the non-military part of the Laboratory. It was a very small airplane. There were about five specialists riding with me and I was the only outsider. They talked shop openly about a topic they called "weapons physics." They were so happily engaged talking about weapons physics that I saw that they very much liked their field. I could see how innocent these people were, thinking about progress in weapons physics. Of course we have similarly dangerous progress in biochemistry—for example regarding viruses. Some dangerous viruses have died out except for a few still kept in a laboratory. So there are many other dangerous elements in science which humanity has learned to cope with—to keep under the rug and under control. This is because our present time is already a little bit advanced on the way towards a global democracy, so people behave as if this democracy were already existent and behave responsibly against this backdrop.

Unfortunately, this rationality has not yet been integrated. Clashes are still possible between powers, and hence there is still the danger of such knowledge getting into hands that are non-benevolent. As soon as one has a society which has the same laws everywhere, humanity can rely on the non-misuse of new results. If you so wish, the danger of misuse is only a consequence of the planet not having been united as of yet, that the United Nations have not enough power. A world government is an option that—like a development that happened in South Africa—can be generated to date on the initiative of a radiant personality who makes it clear to everyone that they should listen to their children. Then this safety problem would go away. Of course, misusable facts could still fall into the hands of desperate individuals. But this risk has always existed for parts of humanity. If the planet had already been united democratically, one could say that Carl-Friedrich von Weizsäcker's advice was sound. And of course, being open in science is one of the oldest traditions of humankind. Dialog is what distinguishes persons from nonpersons, and persons are never quite left alone.

Chapter 25
Open Problems or Eight Dreams of Rationalism

In this session let us conclude the previous 24 sessions with the following question. You have worked and commented on a lot of topics in our previous interviews; do you consider the whole field to be rational, or are there some totally open things that need explanation, too? What is the meaning of rationalism in science?

Thank you for the beautiful question. I would say that science and rationalism are one and the same thing. Science was invented in the modern sense by René Descartes who called his procedure rationalism. Of course, rationalism has a long history; Buddhism and the Abrahamic religions of Judaism, Christianity and Islam are rational edifices of thought. The idea that rationalism is decisive, both for physical survival and for moral survival, is the insight of Descartes. From 1641 until 1927—the advent of quantum mechanics—the rationalist world view of Descartes and his followers loomed large in science as its fundament and core. But this stopped when quantum mechanics was invented. Quantum mechanics causes a big crisis in rationalism. This gloomy mood can be repaired as far as I can tell. Let me go in circles or spirals touching on eight points along the way.

Please start.

As a first unexplained point and inexplicable phenomenon, I would like to mention *COLOR*. You could also take light in general—not the wavelength of the radiation triggering the experience, but the experience itself as a so-called "primary quality." I once met a friend who saw different colors with his two eyes. But this is not the point. Not the relation between colors or their external stimuli is what I have in mind—I am referring here to the self-identity of a tulip's redness, for example. Although it is not very well known, one can prove that color does not exist in science, in nature, anywhere. The same holds true for pain and pleasure, the "talking character" which they all possess. There is nothing in science that is capable of explaining, or predicting, or excusing this primary experience. It is a funnel leading out of the world—or equivalently a "Mehrwert," a value added from outside. What I mean is the "presence" that it represents, the visible or touchable cloth of an angel lowered into our world. My friend Art Winfree, following several rainy weeks in a row, once boarded a little airplane to see the blue sky again, even

A. Sanayei and O. E. Rössler, *Chaotic Harmony*,
Emergence, Complexity and Computation 11, DOI: 10.1007/978-3-319-06781-0_25,

though the machine—offered to the faculty of Purdue University—had a defective altimeter as he realized when mounting the pilot's seat. If the clouds had formed a fog reaching down to the ground on his descent, he would have crashed. He took the chance—this was inexplicable to himself. There is an appeal involved in the "primariness" of color that is plainly irrational. To a scientist, the purity of color is provably nonexistent, its personal touch, its self-identity, its "primariness" is a revelation—a manifest miracle.

People define color based on the wavelength of electromagnetic radiation. What is the problem here when you say it is not rational?

Of course we all know that our retina has three types of receptors to measure the wavelength of light, and that this information is somehow "translated" into color in the brain. But there is no part in the brain, no mechanism, by which redness (say) could be generated. Color is a uniquely subjective experience. As a so-called "quale"—a "how-thing" in Latin—it is not at all represented in the wave length or anywhere else in the circuitry of the brain. There is nothing in nature as seen by science that could explain the transition from the nerve signals, the "pulse coding" or whatever code is used, towards the experience of color as a so-called "primary quality." If one could explain the brain and everything that happens in the brain in terms of pure machine-like dynamical processes and equations—as I would subscribe to—, then one would never even come up with the question of something like color existing. Leibniz saw the problem clearly with his famous example of the windmill in his book "Monadology." He wanted to describe the brain in a scientific, Cartesian sense. The biggest machine of his time was the windmill, it was full of big wheels and little cogwheels; it was the forerunner of the church organ and the computer, built in the Netherlands. He compared the brain with a giant windmill, and there is a little staircase on the side that you can climb-up to enter one of the two side-doors of the windmill which look almost like two "ears" if the turning arms represent a somewhat peculiar analog to a "nose" protruding from this giant "head" erected in the landscape. You can enter the ear of the windmill and describe all the interlocking cogwheels, all their quantitative dynamical relationships. That would be a model of a brain at the time of Leibniz. He continued his tale with the event when the scientist would come out of the windmill's ear door after a long time, standing on top of the staircase with all the news reporters and their cameras (he did not foresee either of these to us familiar realities although he foresaw the telephone and the tape recorder at another occasion) waiting eagerly to get his report.

"Professor Leibniz, have you completely understood the brain?" He would confidently reply; "yes, messieurs, I have completely mapped out and understood it."

"Every detail?"

"Yes."

"How about consciousness?" would be the next question.

"What consciousness?" would then be his reply. Leibniz first saw it clearly after Descartes that the machine theory of the world and the brain does not yield the slightest cue for the existence of anything like color or subjective experience in general. The machine would be totally consistent as a machine, nothing more. Today one would say that there is also a "screen" inside the brain—the "great simulator"—on which everything spatial and short-term temporal is represented in a functionally (if not anatomically) connected fashion. This screen, even though it exists in the brain (and also the brain equation with the big screen or VR machine added), would not have colors and smells on it anywhere, only coded signal sequences talking to each other in coincidences.

So what is your suggestion? You mentioned as your first point the phenomenon of color. Do you have any argument here whether this phenomenon is rational?

It is rational to say that we are confronted here with a miracle, the miracle of the "qualia," the qualities of experience. Philosophers have known this for a long time. Physiologist Emil Du Bois-Reymond in Berlin (whose brother Paul was a mathematics professor in Tübingen) wrote a famous paper in 1872 titled "Ignorabimus" (we shall never know). It dealt exactly with this question. So this is a long-standing problem. People have learned to accept and live with the fact that we shall never know where the qualia come from. In artificial intelligence, the problem is a virtual taboo still. It is the first big riddle in rationalism in my eyes.

What is the second?

The second inexplicable phenomenon in our walking around peripatetically in half-circles is the *NOW*. With the Now it is just the same thing as with the qualia. You can prove that there is no instrument or machine by means of which you could measure when it is Now. Buddha lived only in the Now. The Now again does not exist in science, it is provably non-existent. This is strange since time does exist in physics in many equations and so does simultaneity; even though the latter is observer-specific, as Einstein discovered. But the Now which seems to be moving for us at this very moment is provably nonexistent in science where nothing is moving according to Parmenides. Popper had a long discussion on this with Einstein.

In the Minkowski diagram, the Now part refers to an area in which the speed of light is larger than c, and this is why physicists say it does not exist because it does not make physical sense.

Yes, the Now extends to both sides in between the two light cones, the upwards-opening one of the future and the downwards-opening one of the past. Again, we have a constantly moving-up mid-point in between the two light cones in defiance of the frozen description just given. So the Now itself, which determines the momentary tips of the light cones, has nothing physical behind it that one could tie it to. It is not an element of science.

What is the next "miracle" if any?

The third inexplicable phenomenon is the *ASSIGNMENT* of the body. Why is our consciousness assigned to this particular body? When I was about 3 years old I suddenly realized that the little dimples I knew I had on top of the knuckles on the back of my right hand (i.e. both hands) had disappeared, and I felt it was kind of cruel that I had been changed without being asked beforehand. It was a typical "philosophical" experience. The assignment of the body becomes a problem when one has a health accident—then this prison-type situation fleetingly enters consciousness again. The reason why the body is assigned to a certain consciousness is provably inexplicable. It makes one feel very humble.

A fourth item?

After these three subjectivist miracles—I could have added that the Now is not even inter-subjective according to Roger Shepard—, I come to an indubitably objectivist miracle; the existence of consistent *LAWS* in nature, laws hoped to govern all phenomena. Science has been quite successful finding laws after Descartes had shown how important it is to find consistency as the only way to get rid of the irrationalism of witchcraft, which is the alternative. He was able to prove that *only* if the world is consistent in its quantitative relations, is consciousness acceptable in the sense of it not being a "bad dream." Daniel F. Galoye's 1964 science-fiction novel "Simulacron-Three" comes close: a computer-world in which the "ID-units" find inconsistencies and thereafter get suspicious. That the world be consistent is a necessary condition in order for it not to be a "bad dream" according to Descartes. For it is only under this condition of well-craftedness, that one can prove that one is not in the hands of a malevolent higher-level power. Under this condition, an act of fairness, shown towards a fellow machine-like inhabitant of the universe is bound to put a malevolent demiurge to shame. For the person performing it would be morally superior to the consciousness-giving instance. This offered "chance to get even" for a mere endo observer is the reason why the search for consistency in science is so important. Leibniz called this procedure "Theodicy"—calling the Dream-Giving-Instance before the court of reason. Descartes was apparently sure that heaven would not flunk this test.

A striking example of a previously not seen consistency is the correspondence between symmetry and conservation laws, discovered by Emmy Noether.

Thank you for this illustration. The infinite accuracy of these laws that shows up in her first theorem is fundamental. This accuracy which is transfinitely exact allows for a causal description of thermodynamics, cryodynamics and quantum mechanics. Even though it looks like a huge success, this dream of a consistent causal science is not all that is at stake. For the wonderful machine thereby found could still belong to the Hades. The Hades of Greek mythology (the "shadow world") was in its structure identical to our own upper world—except for the absence of the qualia. The "blood" was missing—so much that the king of the Hades would rather prefer to be the lowliest slave on the surface of the earth bending his back under the whipping of the plough driver as Homer wrote. Even the breath-taking transfinite accuracy of a fractal cannot reach the living color of a quale.

The quantum irreversibility comes to my mind here also. It was recently mentioned by Zeilinger (who was the third of three authors) in a paper in the Proceedings of the Royal Society titled "The Oxford Questions on the foundations of quantum physics." One of the open problems mentioned there was exactly this one. They considered it as an unsolved problem so far.

Which one?

"Is irreversibility fundamental for describing the classical world?" And "how is irreversibility involved in quantum measurement?"

Yes, classical physics is reversible, this reversibility lies at the root of thermodynamics and cryodynamics. Both are reducible to determinism if we assume the number of particles per unit volume in the universe is finite and in addition assume indistinguishability between equal particles. This transfinitely exact, "polished equality" of elementary particles of the same class and the same spin is an intimidating empirical feature of nature. It has the same mathematical rank as fractality. It is so strong in my opinion that it even overrides the current belief in the existence of "primary" statistical phenomena in quantum mechanics. Thermodynamics got interpreted as a statistical theory for a long time, and so did quantum mechanics. However, owing to classical indistinguishability theory, a transfinitely exact deterministic level is bound to underlie both quantum mechanics and classical mechanics. Hermann Weyl knew about these things in the footsteps of Josiah Willard Gibbs, as I learned from my friend Hans Primas.

Next number five?

We now plunge even deeper into non-classical physics. Number five is the hoped-for possibility to arrive at an *EXPLANATION OF h AND c*, Plancks constant and Einstein's constant. Both constants are different from other constants in that they are all-pervasive. In the absence of a rational explanation for both the quantum chancefulness and the observer-centered speed of light, there is a big hole in physics in case its program is still Cartesian rationalism. There exists a single attempt to explain *little h*, that of Otto Sackur of 1913.

We talked about it in a previous session.

Yes—but here is a new point; possibly, *little-c* can be explained along with *little-h*. The Sackur cell in the observer's brain would then be strengthened in its hypothetical fundamental role in physics that we discussed. We had seen that every momentarily closed system possesses a system-specific unit action h^*, including the brain. The brain can be compared to a radio receiver which contains a non-eliminable "noise source" in its input stage. Archimedes was the first to spot an "inevitability" like this when he said "Give me where to stand and I shall move the earth." It was obviously an ironic statement. Being inside a system does not allow you to do this! Maxwell's claim that it is impossible to palpate an atom addresses the same underprivilegedness of a subsystem in a Hamiltonian world. If h is caused by this Maxwellian principle which appears worthy to rule out, then besides h, c can also be seen in this light. A physically defined subsystem in the brain would be responsible for both. The Sackur cell can do this in principle as we saw because it involves micro time reversals t^* which would be "projected-out" into the outside world along with h^*. The micro time reversals along with h^* could then be associated with a spatial length l^*. Then t^* and l^*, taken together, would define an observer-specific velocity, $c^* = l^*/t^*$.

But it would not be "c"?

This absurd speculation can be disproved by calculating. Suppose the unknown speed c^* were numerically close to c. Then the hypothesis of it being an explanation of c could be checked by "calculating back" to obtain the unknown underlying "observer-diameter" l^*. So by inserting the real c, one could determine the unknown l^*. If nonsense comes out, one can forget about this mental adventure. If one puts in the available numbers, a value of about 7.4 μm can be arrived at for l^*, as I once obtained 15 years ago.

This makes no sense.

I agree. To assume that the Sackur cell in the brain had this small spatial size appears even more implausible than was the case when Descartes put the seat of consciousness into the pineal gland. The latter is many thousand times bigger and obviously the wrong candidate. l^* corresponds to the diameter of a single cell in the reticular formation of the brain where consciousness is sometimes assumed to reside. I mentioned all this only because it is clear anyhow that the observer-centeredness of h and c is desperate speculative nonsense. Einstein should rather not have proposed this dangerous experiment which Zeilinger and Pan are preparing now, with the risk that an observer-specific world will be forced upon physics if the two spaceship captains confirm the Bell correlations as no one doubts. I apologize for having alluded to this nightmare.

What is the next point in your "rationalism"?

Number six is the *RELATIVISTIC BELL EXPERIMENT* itself. It will yield demonstrably observer-centered measurements in accordance with endophysics as mentioned. In previous ordinary Bell-Aspect experiments, the correlations seen between both measurements in the experiment were superluminal. One accepted this manifest miracle as a "quantum property" of the physical world. The Copenhagen interpretation allowed for this and since the outcome cannot be misused for sending superluminal messages (only "half-messages"), everyone got used to it. However, the relativistic Bell experiment is a different matter. Since the two measuring stations belong to two mutually receding frames, each measurement is the first in its own frame. In the world of each observer, the observed correlations are compatible with quantum mechanics in the sense of Copenhagen. The two worlds contradict quantum mechanics only if their results are known to all observers. Then, both "momentum" and "position" (or their analogs) have been measured on a pair of correlated particles. Only if each observer lives in a quantum world of his or her own is everything fine again. Then each has reduced the wave-function first, and the other comes later. Our hypothetical causal explanation—that the micro motions inside the observer enter the result—then is the only remaining way to interpret the relativistic Bell experiment. This makes the latter maximally important.

So far, the two interpretations, Copenhagen and Everett, cannot be experimentally distinguished.

This is the point: now they suddenly can. Einstein no doubt was aware of this option. The Zeilinger quantum satellite experiment to check on the latter has been in progress since 2001. It will prove that the assignment of an Everett world is an empirical reality in physics—unless the Bell correlations are violated!

Your seventh point?

We enter ordinary classical intuition again with the "seventh rationally to accept miracle," which is *EXTERIORITY*. The term is due to French philosopher Emmanuel Levinas. It refers to the fact that there exists an infinitely strong influence, not only "vertically" (with the quantum assignment conditions of the sixth point), but also "horizontally." We find ourselves with our bodies "outside" (exterior) to our fellow neighbors. Being outside another person means that you can do good, or refrain from doing good. The infinite power of exteriority stems from being not identical with the other. For the consciousness-endowed person is not a mere machine since consciousness is an add-on, an infinite privilege as Descartes saw. Levinas quotes Jesus for having made the same discovery of an infinite power existing here, with his notion of the "neighbor." The infinite privilege turns you, the responsibility-carrying conscious subject, into a "hostage"—a strangely moving idea. Levinas almost wails over how much pity he deserves for being so infinitely powerful. By being on the outside of the other person, in a rational Cartesian machine world, a person endowed with consciousness is in a most pitiful situation. I once listened to Levinas on TV, he was a very old man at the time and one of the most charming souls I ever encountered. The same exteriority principle is by the way implemented in "*Lampsacus,*" hometown of all persons on the Internet, proposed in 1994.

One more open problem?

The eighth Cartesian point would be the *ORIGIN OF MUTUAL TRUST* in the life of the individual as an acquired non-genetic trait based on the smile feedback. I just realize I can use this last point to come back to your beauty question in a kind of summary: There would exist three types of beauty in science. The first is the beauty of the consistency of the laws (Descartes-Newton-Einstein-Dirac). The second is the beauty of the smile-feedback (Darwin-van Hoof-Disney-Buber). The third stands behind both as the substance injected into the assigned Now captured in an ancient Egyptian phrase ("heaven and earth are filled with your splendor").

Thank you for this futuristic conclusion.

Bibliography

Abraham, M. (1914). More recent gravitation theories (in German). *Jahrbuch für Radioaktivität und Elektronik, 11*, 470–520.
Anaxagoras (500–428 B.C.), Fragment # 12. In G. S. Kirk, & Raven, J. E. (Eds.), *The presocratic philosophers* (pp. 372–374). London: Cambridge University Press 1957, 1979.
Anaximander (610–546 B.C.). In Kirk and Raven, *The prescoratic philosophers* (pp. 110–113).
Anderton, R., & Nedelkovich, D. (2013). *Natural Philosophy and Relativity of Boscovich*. Lulu.
Aristarchus of Samos (310–230 B.C.). See Archimedes (ca. 287–212 B.C.), *The Sand Reckoner*. [Quote: "Aristarchus' hypotheses are that the fixed stars and the Sun remain unmoved, that the Earth revolves about the Sun on the circumference of a circle, the Sun lying in the middle of the orbit, and that the sphere of fixed stars, situated about the same center as the Sun, is so great that the circle in which he supposes the Earth to revolve bears such a proportion to the distance of the fixed stars as the center of the sphere bears to its surface."].
Aspect, A. (1999). Bell's inequality test: More ideal than ever. *Nature, 398*, 189–90.
Avicenna, M. E. (2005). *The metaphysics of the healing*. In M. E. Marmura (Ed.), Utah: Brigham Young University.
Bacon, F. (1960). Novum organum scientiarum. 1620. In F. H. Anderson (Ed.) *The new organum and related writings*. New York: Liberal Arts Press.
Baier, G., & Klein, M. (Eds.). (1991). *A chaotic hierarchy*. Singapore: World Scientific.
Bateson, G. (1972). *Steps to an ecology of mind*. New York: Ballantine.
Bell, J. S. (1964). On the Einstein-Podolsky-Rosen. *Physics, 1*, 195–200; reprinted in: Bell, J. S. (1987). *Speakable and unspeakable in quantum mechanics*. Cambridge: Cambridge University Press.
Bell, J. S. (1981). Quantum mechanics for cosmologists. In C. Isham, R. Penrose, & D. Sciama (Eds.), *Quantum gravity* (Vol. 2, pp. 611–637). Oxford: Clarendon Press. reprinted in: Bell, J. S. (1987). *Speakable and unspeakable in quantum mechanics*. Cambridge University Press.
Bendixson, I. (1901). Sur les courbes définies par des équations différentielles. *Acta Mathematica, 24*, 1–88.
von Bertalanffy, L. (1932). *Theoretical biology (Theoretische Biologie)*. Stuttgart: Borntraeger.
von Bertalanffy, L. (1953). *Biophysics of the equilibrium-of-flow (Biophysik des Fliessgleichgewichts)*. Braunschweig: Vieweg.
Binney, J., & Tremaine, S. (2008). *Galactic dynamics* (2nd edn). Princeton: Princeton University Press.
Bohm, D. (1951). *Quantum theory*. Englewood Cliffs: Prentice-Hall.
Bohr, N. (1913). On the constitution of atoms and molecules. *Philosophical Magazine, 26*, 4.
Bohr, N. (1935). Can quantum mechanical description of physical reality be considered complete? *Physical Review, 48*, 696–702.
Boltzmann, L. (1964). *Lectures on gas theory* (Vorlesungen über Gastheorie, Vol. 1, Barth, 1896). Leipzig: University of California Press.
Boscovich, R. J. (1758). *A theory of natural philosophy* (Theoria philosophiae naturalis). Vienna.

A. Sanayei and O. E. Rössler, *Chaotic Harmony*,
Emergence, Complexity and Computation 11, DOI: 10.1007/978-3-319-06781-0,

Briggs, G. A. D., Butterfield, J. N., & Zeilinger, A. (2013). The Oxford questions on the foundations of quantum physics. *Proceedings of the Royal Society A, 469*, 1–8.
Boer, H. (2004). KAM theory—the legacy of Kolmogorov's 1954 paper. *Bulletin of the American Mathematical Society, 41*, 507–521.
Bruno, G. (1584). *De l'infinito universo e mondi*. Venice 1584. (On the infinite universe and worlds.) Reprint Venize 2012.
Buber, M. (1958). *I and Thou*. New York: Scribner.
Bunimovich, L. A. (2000). Billiards and other hyperbolic systems with singularities. In Ya G Sinai (Ed.), *Dynamical systems, ergodic theory and applications* (pp. 192–234). Berlin: Springer.
Buridan's ass (2009). In: *Oxford companion to phrase and fable*. Encyclopedia.com.
Calaprice, A. (Ed.). (2002). *Dear professor Einstein—Albert Einstein's letters to and from children (foreword by Evelyn Einstein, essay by Robert Schulman)*. New York: Prometheus Books.
Cantor, G. (1874). On a property of the intrinsic idea of all real algebraic numbers (in German). *Journal für die Reine und Angewandte Mathematik, 77*, 258–262.
Chandrasekhar, S. (1943). Dynamical friction I. *Astrophysical Journal, 97*, 1–27.
Charnov, E. L. (1976). Optimal foraging: The marginal value theorem. *Theoretical Population Biology, 9*, 129–136.
Clausius, R. (1865). On different easy to apply versions of the principal equations of the mechanical theory of heat (in German). *Annalen der Physik und Chemie, 125*, 353–390.
Cook, R. J. (2009). Gravitational space dilation, arXiv:0902.2811.
Copernicus, N., *On the Revolutions of the Celestial Spheres* (De revolutionibus orbium coelestium). 1st edn. Nuremberg 1543, 2nd edn. Basel 1566.
Damour, T. (1987). The problem of motion in Newtonian and Einsteinian theory. In S. Hawking, & W. Israel (Eds.), *300 years of gravitation* (pp. 128–198). Cambridge: Cambridge University Press.
Darvas, G. (2012). Isotopic field charge—spin conservation in general relativity theory. In M. C. Duffy, V. O. Gladyshev, A. N. Morozov, P. Rowlands (Eds.), *Physical interpretations of relativity theory* (pp. 53–65). Liverpool: Sutherland.
Darwin, C. (1859). *On the origin of species by means of natural selection—or the preservation of favoured races in the struggle for life*. London: Murray.
Darwin, C. (1872). *The expression of the emotions in man and animals*. London: Murray.
Dehmelt, H., van Dyck, R. S., Schwinberg, P. B., Gabrielse, G., & Bull, G. (1979). Single elementary particle at rest in free space I-IV. *American Physical Society, 24*, 757.
Descartes, R. (1641). *Meditations on the first philosophy* (Meditationes de prima philosophia) also known as *Metaphysical Meditations*.
Devaney, R. L. (1996). *The fractal geometry of the Mandelbrot set*. Paperback. Berkeley: Key curriculum Press.
de Waal, F. (2006). In S. Macedo, & J. Ober (Eds.) *Primates and philosophers: How morality evolved*. Princeton: Princeton Science Library.
Diebner, H. H., & Rössler, O. E. (1998). A deterministic entropy based on the instantaneous phase space volume. *Zeitschrift fur Naturforschung, 53a*, 51–60.
Dirac, P. A. M. (1926). On the theory of quantum mechanics. *Proceedings of the Royal Society A, 112*, 661–677.
Dirac, P. A. M. (1937). The cosmological constants. *Nature, 139*, 323.
Dirac, P. A. M. (1958). *The principles of quantum mechanics* (4th ed.). London: Oxford University Press.
Du Bois-Reymond, E. (1912). *On the limits of recognizing nature* (Über die Grenzen des Naturerkennens) 1872; Lectures Ignorabimus *by* Emil du Bois-Reymond (edited by Estelle du Bois-Reymond), (pp. 441–473). Leipzig: Veit.
Eddington, A. S. (1981). *The nature of the physical world*. Gifford lectures 1928. MacMillan: University of Michigan Edition.

Ehrenfest, P. (1911). Which traits of the light quantum hypothesis play an essential role in the theory of heat radiation? (in German). *Annalen der Physik, 341*, 91–118.
Einstein, A. (1905) On the electrodynamics of moving bodies (in German). *Annalen der Physik, 17*, 891–921, In *The principles of relativity*. London: Methuen (1923).
Einstein, A. (1907). On the relativity principle and the conclusions drawn from it (in German). *Jahrbuch der Radioaktivität, 4*, 411–462.
Einstein, A., Podolsky, B., & Rosen, N. (1935). Can quantum-mechanical description of physical reality be considered complete? *Physical Review, 47*, 777–780.
Eliasson, O. (2003). In S. May (Ed.) *The weather project*. London: Tate.
Eliasson, O. (Ed.). (2010). *Otto's impossible talks*. Inst. Raumexperimente Berlin, Verlag der Universität der Künste Berlin.
Engelmann, W. (2007). Rhythms of life—an introduction using selected topics and examples. http://bioclox.bot.biologie.uni-tuebingen.de/.
Everett, H. (1957). "Relative State" formulation of quantum mechanics. *Reviews of Modern Physics, 29*, 454–462.
Farmelo, G. (Ed.). (2002). *It must be beautiful: Great equations of modern science*. London: Granta Books.
Farmelo, G. (2009). *The strangest man—the hidden life of Paul Dirac, quantum genius*. London: Faber and Faber.
Finkelstein, D. (1959). Some new conservation laws. *Annals of Physics, 6*, 230–243.
Fischer, J., & Plessner, H. (2013). Laughter and smile—an investigation of the limits of human behavior (in German). In *Hauptwerke der Emotionssoziologie* (pp. 274–279). Berlin: Springer.
Fischer, U. R., & Schopohl, N. (2001). Hal l state quantization in a rotating frame. *Europhysics Letters, 54*, 502–507.
Flamm, L. (1916). Contributions to Einstein's gravitation theory (in German). *Physikalische Zeitschrift, 17*, 448–454.
Frayn, M. (1998). *Copenhagen*. London: Methuen.
Forward, R. F. (1980). *Dragon's egg*. New York: Ballatine Books.
Fournier d'Albe, E. E. (1907). *Two new worlds: I. The infra world: II. The supra world.* London: Longmans and Green.
Frauendiener, J., Hoenselaers, C., & Konrad, W. (1999). A shell around a black hole. *Classical and Quantum Gravity*, **7**, 585–591.
Fredkin, E., & Toffoli, T. (1982). Conservative logic. *International Journal of Theoretical Physics, 21*, 219–253.
Freely, J. (2009). *Aladdin's lamp—how Greek science came to Europe through the Islamic world.* New York: Vintage Books.
Friedmann, A. (1999). On the possibility of a world with constant negative curvature of space (in German). *Zeitschrift für* Physik *21*, 326–332 (1924). (English translation in *General Relativity and Gravitation 31*, 2001–2008, 1999).
Froebel, F. (1844). In Blow S. E. (Ed.), *Mother's songs, games and stories: Froebel's the Kose-Lieder (Mutter- und Koselieder)*. Bad Blankenburg.
Galilei, G. (1914). Discorsi e dimostrazioni matematiche intorno a due nuove scienze. Leiden 1638. In *Discourses and Mathematical Demonstrations Relating to Two New Sciences*. New York: Dover.
Galouye, D. F. (1964). *Simulacron-3*. New York: Bantam Books. (*Welt am Draht* [A Puppeteer's World], Munich: Goldmann, 1965).
Ganoczy, A. (1995). *Chaos, chance, belief in creation—Chaos theory as a challenge for theology* (in German). Mainz: Matthias-Grünewald-Verlag.
Garey, M. R., & Johnson, D. S. (1979). *Computers and intractability—a guide to the theory of NP-completeness*. San Francisco: Freeman.

Gibbs, J. W. (1960). *Elementary principles in statistical mechanics, developed with especial reference to the rational foundation of thermodynamics*. New York: Scribner 1902 (New York: Dover 1960).

Gibson, W. (1984). *Neuromancer*. New York: Ace Books.

Giel, K. (1988). *Aspects of the dialogical—on the philosophical pedagogics of Martin Buber* (in German). http://www.klaus-giel.de/doc/Buber.pdf.

Gödel, K. (1931). Über formal unentscheidbare Sätze der Principia Mathematica und verwandter Systeme I. *Monatshefte für Mathematik und Physik 38*, 173–198 (1931) (*On formally undecidable propositions of principia mathematica and related systems I*. New York: Basic Books, 1963; New York: Dover, 1992).

Goldman, M. (1983). *The demon in the Aether—the story of James Clerk Maxwell*. Edinburgh: Paul Harris.

Gregory of Nazianzus (1975). *On human nature* (De humana natura). In J. Plevnic (Ed.), J. Egan, *Gregory of Nazianzus and the logos doctrine* (pp. 281–322). ON: Willowdale.

Gumowski, I., & Mira, C. (1980). *Recurrences and discrete dynamic systems*. Berlin: Springer.

Haken, H. (1970). Laser theory. In S. Flügge (Ed.), *Handbuch der Physik* (Vol. 25/2c). Berlin: Springer.

Hahn, O. (1944). *Artificial transmutations of atoms and the splitting of heavy nuclei* (in German). Publications of the German Scientific Institute (DWI), Stockholm, Series 3: Naturwissenschaften, No. 1. Stockholm: Almquist and Wiksells.

Hawking, S. W. (1974). Black hole explosions? *Nature*, *248*(5443), 30–31.

Heaviside, O. (1892). On the forces, stresses, and fluxes of energy in the electromagnetic field. *Philosophical Transactions of the Royal Society, London*, *183*, 423–480.

Heraclitus (2001). Fragment # 70: Metabolizing it rests (Metabállon anapaúetai). In *Fragments—the collected wisdom of Heraclitus*. New York: Haxton Brooks Viking.

Higgs boson (14 March 2013). New results indicate that particle discovered at CERN is a Higgs boson. *CERN-Press release*.

Hilbert, D. (1902). Mathematical problems. *The Bulletin of the American Mathematical Society*, *8*, 437–479.

Hill, G. W. (1878). Researches in the lunar theory. *The American Journal of Mathematics*, *1*, 5–26, 129–147, 245–260.

Hoppenstaedt, F. (Ed.) (1979). *Nonlinear oscillations in biology* (*Lectures in Applied Mathematics*, *17*) Rhode Island: The American Mathematical Society.

Hoyle, F. (1950). *The nature of the universe—a series of broadcast lectures*. Oxford: Basil Blackwell.

Hoyle, F. (1957). *The black cloud*. London: William Heinemann.

Hubble, E. (1929). A relation between distance and radial velocity among extra-galactic nebulae. *Proceedings of the National Academy of Sciences*, *15*, 168.

Hulse, R. A., & Taylor, J. H. (1975). Discovery of a pulsar in a binary system. *Astrophysical Journal*, *195*(pt. 2), L51–L53.

Huxley, J. (1914). The courtship habits of the Great Crested Grebe (Podiceps cristatus), with an addition to the theory of sexual selection. *Proceedings of the Zoological Society of London*, *84*, 491–562.

Jordan, P., von Neumann, J., & Wigner, E. (1934). On an algebraic generalization of the quantum mechanical formalism. *Annals of Mathematics, 35*, 29–64.

Julia, G. M. (1918). Mémoire sur l'itération des fonctions rationnelles. *Journal de Mathématiques Pures et Appliquées*, *8*, 47–246.

Kant, I., *Critique of pure reason* (1781). In P. Guyer & A. Wood, *Cambridge edition of the works of Immanuel Kant*. Cambridge: Cambridge University Press 1999.

Kaplan, J. L., & Yorke, J. A. (1979). Chaotic behavior of multidimensional difference equations. In H. -O. Peitgen & H. -O. Walther (Eds.), *Functional differential equations and approximations of fixed points: Proceedings*, Bonn, July 1978 (pp. 204–227). Berlin: Springer.

Kauffman, S. (1996). *At home in the universe—the search for the laws of self-organization and complexity*. Oxford: Oxford University Press.
Kaufmann, W. (1903). On the "electromagnetic mass" of the electrons (in German). *Göttinger Nachrichten*, *3*, 90–103.
Kennefick, D. (2005). Einstein versus the physical review. *Physics Today*, *58*, 43–48.
Kepler, J. (1609). *New astronomy*. Cambridge: Cambridge University Press 1993.
Kloeden, P., Rossler, O. E., & Rossler, R. (1990). A predictive model for life expectancy curves. *BioSystems*, *24*, 119–125.
Kocher, C. A., & Commins, E. D. (1967). Polarization correlation of photons emitted in an atomic cascade. *Physical Review Letters*, *18*, 575–577.
Köhler, W. (1921). *Intelligence tests in apes* (*in German*). Berlin: Springer 1963.
Korzybski, A. (1921). *Manhood of Humanity*. New York: Dutton.
Kossel, W. (1919). On the physical nature of the valence forces (in German). *Naturwissenschaften*, *7*(336–345), 360–366.
Kozak, J.J., & Rossler, O. E. (1982). Weak mixing in a quantum system. *Z. Naturforsch. 37a*, 33–38.
Kümmel, F. (1972). The consciousness of time as the unification of spontaneity and receptivity—On the Foundation of the objectivity of understanding in the Critique of Pure Reason (in German). *Zeitschrift für Philosophische Forschung*, *26*(1), 21–28.
Lakatos, I. (1976). *Proofs and refutations—the logic of mathematical discovery*. Cambridge: Cambridge University Press.
Lakatos, I. (1978). *Mathematics, science and epistemology. Philosophical papers* (Vol. 2). Cambridge: Cambridge University Press.
Landau, L. D. (1965). In D. ter Haar (Ed.), *Collected papers of L.D. Landau*, 1930. London: Gordon and Breach.
Landauer, R. (1978). Stability in the dissipative steady state. *Physics Today*, *31*, 23–24.
Langevin, P. (1911). The evolution of space and time (in French). *Scientia 10*, 31–54 (1973 transl. J. B. Sykes; also in Wikisource).
Lasswitz, K. (1901). *The universal library* (in German) [English Translation in C. Fadiman, (Ed.), *Fantasia mathematica*. New York: Simon and Schuster 1958].
Leibniz, G. W., & Clarke, S. (1716). R. Ariew (Ed.), *The Leibniz-Clarke Correspondence*. Indianapolis/Cambridge: Hackett 2000.
Leibniz, G. W. (1714). *Monadology* (La Monadologie) (English translation by Robert Latta, 1898, George MacDonald Ross, Leeds 1999).
Lemaître, G. (1927). A homogeneous universe of constant mass and crossing light rays taking into account the radial velocity of extra-galactic nebulae (in French). *Annales de la Société Scientifique de Bruxelles A*, *47*, 49–56.
Lennard-Jones, J. E. (1924). On the determination of molecular fields. *Proceedings of the Royal Society of London, Series A*, *106*, 463–477.
Letellier, C. (2013). *Chaos in nature*. Singapore: World Scientific.
Levin, K. (1936). *Principles of topological psychology*. New York: McGraw-Hill.
Lévinas, E. (1947). *Time and the otherperson* (Le temps et l'autre) (English Translation). Pittburgh: Duquesne University Press 1990.
Li, T.-Y., & Yorke, J. A. (1975). Period three implies chaos. *The American Mathematical Monthly*, *82*, 985–989.
Liénard, A. (1928). Etude des oscillations entretenues. *Revue générale de l'électricité*, *23*, 901–912, 946–954.
Linke, D. B. (2003). *Media theory, brain research and accepting Turkey into the EU—international Flusser Lecture* (in German). Cologne: Verlag der Buchhandlung Walter König.
Lorenz, E. N. (1963). Deterministic nonperiodic flow. *Journal of the Atmospheric Sciences*, *20*, 130–141.
Lorenz, K. (1954). *Man Meets Dog*. London: Routledge Classics.

Lorenz, K. (1978). *Behind the mirror*. Engelska: Mariner Books. (Die Rückseite des Spiegels [The Mirror's Flip Side]. Munich: Piper, 1973).
Loschmidt, J. J. (1866). On the size of the molecules of the air (in German). *Sitzungsberichte der kaiserlichen Akademie der Wissenschaften Wien 52*, Abt. II, 395–413.
Loyola, I. (1997). cf. *Life of St. Ignatius of Loyola*. Charlotte: TAN Books.
Lullies, H. (1958). *Physiologie*, vols. 1 and 2. Cologne, Tropon.
Lyman, T. (1906). The spectrum of hydrogen in the region of extremely short wave-length. *Memoirs of the American Academy of Arts and Sciences, 13*, 125–146.
Lynden-Bell, D. (1996). Inertia. In O. Lahav, E. Terlevich, & R. J. Terlevich (Eds.), *Gravitational dynamics: Proceedings of the 36th Herstmonceaux Conference in honour of Professor D. Lynden-Bell's 60th birthday* (pp. 235–243). Cambridge: Cambridge University Press.
Maimonides, M. (1963). *A guide for the perplexed* (transl. by S. Pines). University of Chicago Press.
Malina, R. F. (2 Jan 2012). A role for the sublime in art/science? *Leonardo Quarterly Reviews*, pp. 14–19.
Mandelbrot, B. B. (1977). *The fractal geometry of nature*. New York: Freeman.
Maxwell, J. C. (2001). *Theory of heat*, 2nd edn., 1872. New York: Dover.
May, R. M. (1974). Biological Populations with nonoverlapping generations: Stable points, stable cycles, and chaos. *Science, 186*, 645–647.
Mayer-Kress, G. (1999). In R. Stettler (Ed.), *3rd International Symposium of Science, Technics, and Aesthetics: Frontier Communication: Human Beings, Apes, Whales, Electronic Networks*, Lucerne, Switzerland.
Meitner, L., & Frisch, O. R. (1939). Disintegration of uranium by neutrons: A new type of nuclear reaction. *Nature, 143*, 239–240.
Menger, K. (2002). *Selecta Mathematica* (Vol. 1). Vienna: Springer-Wien.
Michelson, A. A., & Morley, E. W. (1886). Influence of motion of the medium on the velocity of light. *American Journal of Science, 31*, 377–386.
Milnor, J. (1985). On the concept of attractor. *Communications in Mathematical Physics, 99*, 177–195.
Minkowski, H. (1909). Space and time (Raum und Zeit). *Physikalische Zeitschrift 10*, 104–111 [for Minkowski diagram, see: N. D. Mermin, *Space and time in special relativity* (pp. 155–199). New York: McGraw-Hill 1968].
Misner, C. W., Thorne, K. S., & Wheeler, J. A. (1973). *Gravitation*. San Francisco: Freeman.
Moser, J. (1962). On invariant curves of area-preserving mappings of an annulus. *Nachr. Akad. Wiss. Göttingen Math.-Phys. Kl. II*, 1–20.
Mössbauer, R. L. (1958). Nuclear resonance fluorescence by gamma radiation in Ir-191 (in German). *Zeitschrift für Physik A, 151*, 124–143.
Newton, I. (1687). *Mathematical principles of natural philosophy* (Philosophiae naturalis principia mathematica) ("The Principia", transl. Andrew Motte, 1729).
Nitschke, A. (1962). *The orphaned child of nature* (in German). Tübingen: Niemeyer.
Noether, E. (1918). Invariant variantional problems (in German). *Nachr. Königl. Gesellschaft der Wissenschaften Göttingen, Math-phys. Klasse,* 235–257.
Olbers, H. W. (1758–1840, rediscoverer of Kepler's 1610 result), see Mandelbrot, B. *The fractal geometry of nature* (p. 103). New York: Freeman, 1977.
Olds, J. (1976). Reward and drive neurons. In: A. Wauquier & E. T. Rolls (Eds.), *Brain stimulation reward* (pp. 1–30). Amsterdam: North-Holland.
Oppenheimer, R., & Snyder, H. (1939). On continued gravitational contraction. *Physical Review, 56*, 455–459.
Ord, G. (2005). *Transfinite physics*. Amsterdam: Elsevier.
Pais, A. (1982). *Subtle is the Lord—the science and life of Albert Einstein*. Oxford University Press.

Pan, J.-W., Simon, C., Brukner, C., & Zeilinger, A. (2001). Entanglement purification for quantum communication. *Nature, 410*, 1067–1070.
Pauli, W. (1924). Discovery of nulear spin to explain the hyperfine structure of the atomic spectra (in German). *Naturwissenschaften 12*, 74 [Cf. Pauli, W. In Kronig and Weisskopf (Eds.), *Collected works in two volumes*. New York: Wiley, 1964].
Penrose, R. (1989). *The emperor's new mind—concerning computers, minds, and the laws of physics* (p. 287, Fig. 6.32). Oxford University Press.
Perlmutter, S., & Schmidt, B. P. (2003). Measuring cosmology with supernovae. *Lecture Notes of Physics, 598*, 195–217.
Pickover, C. A. (2009). *The math book—from Pythagoras to the 57th dimension—250 milestones in the history of mathematics*. New York: Sterling.
Plessner, H. (1950). The smile (in German). In: *Pro regno, pro sanctuario, Festschr. for G. van der Leeuw*.
Poincaré, H. (1993). *New methods of celestial mechanics* (3 vols, in French 1892). College Park: American Institute of Physics.
Poincaré, H. (1958). *The value of science* (in French, Flammarion, Paris 1904). New York: Dover.
Popper, K. R. (2002). *The logic of scientific discovery (Logik der Forschung, 1934)*. New York: Routledge Classics.
Popper, K. R. (1979). *Objective knowledge—an evolutionary approach* (revised edn.). New York: Oxford University Press.
Pound, R. V., & Rebka, G. A. (1960). Apparent weight of photons. *Physical Review Letters, 4*, 337–341.
Pöschel, J. (2001). A lecture on the classical KAM-theorem. *Proceedings of Symposia in Pure Mathematics (AMS), 69*, 707–732.
Prehn, H. (2001). The neuro-loop—biofeedback interfaces. In H. H. Diebner, T. Druckrey & P. Weibel (Eds.), *Sciences of the interface—Festschrift for Otto E. Rössler* (pp. 202–231). Tübingen: Genista-Verlag.
Prigogine, I. (1955). *Introduction to thermodynamics of irreversible processes*. Springfield: C.C. Thoma.
Primas, H. (2003). Between mind and matter. *Mind and Matter, 1*, 81–119.
Putman, H. (1981). *Reason, truth, and history*. Cambridge: Cambridge University Press.
Rashevsky, N. (1954). Topology and life: In search of general mathematical principles in biology and sociology. *Bulletin of Mathematical Biophysics, 16*, 317–348.
Riess, A., et al. (1998). Observational evidence from supernovae for an accelerating universe and a cosmological constant. *Astronomical Journal, 116*, 1009–1038.
Rindler, W. (2001). *Relativity—special, general, and cosmological*. New York: Oxford University Press.
Rizzolatti, G., & Craighero, L. (2004). The mirror-neuron system. *Annual Review of Neuroscience, 27*, 169–192.
Rosen, R. (1967). *Optimality principles in Biology*. London: Butterworth.
Rosen, R. (1970). *Dynamical system theory in biology*. New York: Wiley.
Ross, J., Corlan, A. D., & Müller, S. C. (2012). Proposed principles of maximum local entropy production. *Journal of Physical Chemistry B, 116*, 7858–7865.
Rössler, O. E. (1971). A system-theoretic model of biogenesis (in German). *Zeitschrift für Naturforschung, 26 b*, 741–746.
Rössler, O. E. (1974). Adequate locomotion strategies for an abstract organism in an abstract environment: A relational approach to brain function. *Lecture Notes in Biomathematics, 4*, 342–369.
Rössler, O. E. (1975). Mathematical model of a proposed treatment of early infantile autism. *San Diego Biomedical Symposium, 14*, 105–110.
Rössler, O. E. (1976). An equation for continuous chaos. *Physics Letters A, 57*, 397–398.
Rossler, O. E. (1976/1979). *Chaos Movie*. www.youtube.com/watch?v=Tmmdg2P1RIM

Rössler, O. E. (1978). Deductive biology—some cautious steps. *Bulletin of Mathmatics, 40*, 45–58.
Rössler, O. E. (1979). Chaotic oscillations: An example of hyperchaos. In F. Hoppenstaedt (Ed.), *Nonlinear oscillations in biology* (Vol. 17, pp. 141–156). Lectures in applied mathematics Rhode Island: American Mathematics Society.
Rössler, O. E. (1979). An equation for hyperchaos. *Physics Letters A, 71*, 155–157.
Rössler, O. E. (1979). Chaos. In W. Güttinger & H. Eikemeier (Eds.), *Structural stability in physics* (pp. 290–309), Berlin: Springer.
Rössler, O. E. (1992). Bell's symmetry. *Symmetry: Culture & Science*, *3*(4), 385–400.
Rössler, O. E. (1994). Micro-constructivism. *Physica D, 75*, 438–448.
Rössler, O. E. (1996). Jumping identities of particles. *Symmetry: Culture & Science*, *7*, 307–319.
Rössler, O. E. (1998). *Endophysics—the world as an interface*. Singapore: World Scientific.
Rössler, O. E. (2000). Einstein's Aarau vision. In: J. Fischer (Ed.), *Art Space Aarau* (pp. 191–194). Aarau: Verlag Vittorio de Angelis.
Rossler, O. E. (2006). Static cosmology from chaos-borne Hubble law. *Nonlinear Phenomena in Complex Systems*, *9*, 53–60.
Rössler, O. E. (2011). Variantology: Einstein-Bohr battle confirms Everett's eternal now. In Siegfried Zielinski, Eckhard Fürlus & Daniel Irrgang (Eds.), *Variantology 5: Neapolitan affairs, on deep time relations of arts, sciences and technologies* (pp. 417–432). Verlag der Buchhandlung Walter Köenig: Cologne.
Rössler, O. E. (2011). The new science of cryodynamics and its connection to cosmology. *Complex Systems*, *20*, 105–113.
Rössler, O. E. (2012). Einstein's equivalence principle has three further implications besides affecting time: T-L-M-Ch theorem ("Telemach"). *African Journal of Mathematics and Computer Science Research*, *5*, 44–47.
Rossler, O. E. (2013). Olemach theorem: Angular-momentum conservation implies gravitational redshift proportional change of length, mass and charge. *European Scientific Journal*, *9*(6), 38–45.
Rossler, O. E. (2013). Rolling ball in breathing plane-tree alley paradigm. *European Scientific Journal*, *9*(27), 1–7.
Rössler, O. E. (2014). *The brain equation. In: ISCS 2013: Interdisciplinary Symposium on Complex Systems*, (pp. 47–56). Berlin: Springer.
Rossler, O. E., Rossler, R., & Landahl, H. D. (1978). Arrhythmia in a periodically forced excitable system. In: *Sixth International Biophysics Congress Kyoto*, abstracts (p. 296). International Union for Pure and Applied Biophysics.
Rössler, O. E., Hudson, J. L., & Klein, M. (1989). Chaotic forcing generates wrinkled boundary. *Journal of Physical Chemistry*, *93*, 2858–2860.
Rössler, O. E., Kampis, G., Nadler, W., Musterle, W., Schapiro, B., & Urban, B. (1990). Highly parallel implementation of an autonomous direction optimizer with cognition (abstract). *Biophysical Journal*, *57*, 194a; contains proposal for the *Pandaka pygmaea Brain Research Institute*.
Rössler, O. E., & Hudson, J. L. (1993). A "superfat" attractor with a singular-continuous 2-D Weierstrass function in a cross section. *Zeitschrift für Naturforschung, 48a*, 673–678.
Rossler, O. E., & Hartmann, G. C. (1995). Attractors with flares. *Fractals*, *3*, 285–296.
Rössler, R., & Rössler, O. E. (Eds.). (1994). *Jonas' World - A Child's Thoughts* (in German). Reinbek: Rowohlt.
Rössler, O. E., & Conrad, M. (2002). Interface theory and almost-solipsism. *Journal of New Energy*, *6*, 120–124.
Rössler, O. E., Fröhlich, D., & Kleiner, N. (2003). A time-symmetric Hubble-like law: Light rays grazing randomly moving galaxies show distance-proportional redshift. *Zeitschrift für Naturforschung*, *58a*, 807–809.

Rössler, O. E., Sanayei, A., & Zelinka, I. (2013). Is hot fusion made feasible by the discovery of cryodynamics? In: I. Zelinka, O. E. Rössler, V. Snásel, A. Abraham & E. Corchado (Eds.), *Nostradamus—modern methods of prediction, modeling and analysis of nonlinear systems* (pp. 1–4). Berlin: Springer.

Rossler, R., Götz, F., & Rossler, O. E. (1979). Chaos in endocrinology. *Biophysical Journal, 25*, 216a.

Sackur, O. (1913). The universal significance of the so-called elementary quantum of action (in German). *Annalen der Physik, 40*, 67–86.

Sanayei, A. (2013). Revisiting Dirac and Schrödinger: A proof offered for the non-relativistic time-dependent Schrödinger equation. arXiv:1309.1787.

Scheler, M. Die Stellung des Menschen im Kosmos. Bouvier, Bonn 1926 (17th edn. 2007), Man's Place in Nature. Noonday, New York 1961.

Schiller, F., *On the Aesthetic Education of Man* (in German, 1795), transl. by E.M. Wilkinson and L.A. Willoughby. Oxford: Clarendon Press 1967.

Schrödinger, E. (1926). An undulatory theory of the mechanics of atoms and molecules. *Physical Review*, *28*, 1049–1070.

Schrödinger, E., Die gegenwärtige Situation in der Quantenmechanik. *Naturwissenschaften*, *23*, 844–849 (1935); Present situation in quantum mechanics. *American Philosophical Society*, *124*, 323–338 (1980).

Schuster, G., Smits, W., & Ullal, J. (2008). *Thinkers of the jungle—the orangutan report*. Potsdam: Ullmann.

Seaman, B., & Rossler, O. E. (2011). *Neosentience—the benevolence engine*. Chicago: Intellect.

Shepard, R. N. (1995). Personal communication.

Sinai, Ya. G. (1970) Dynamical systems with elastic reflections. *RussianMathematicalSurveys*, *25*, 137–189.

Sinai, Ya. G. (2004) What is a billiard? *Notices* of the *American Mathematical Society, 51*, 412–413.

Slipher, V. M. (1917). Nebulae. *Proceedings of the American Philosophical Society*, *56*, 403–409.

Smale, S. (1967). Differentiable dynamical systems. *Bulletin of the American Mathematical Society*, *73*, 747–817.

Smoluchowski, M. (1906). On the kinetic theory of the Brownian molecular motion and the suspensions (in German). *Annalen der Physik*, *326*, 756–780.

Sonnleitner, K. (2010). *StV4: A symplectic time-reversible Störmer-Verlet procedure of the fourth order for Hamiltonian multi-particle systems, with two applied examples (gas, T-tube configuration)*, in German. Ph.D. Thesis, University of Tübingen.

Spaemann, R. (1996). *Persons—essays on the difference between "something" and "someone"* (in German). Stuttgart: Klett-Cotta.

Spinoza, B. (1998). *Principia philosophiae cartesianae 1663. The Principles of Cartesian Philosophy*. Indianapolis: Hackett.

Stone, A. Douglas (2005). Einstein's unknown insight and the problem of quantizing chaos. *Physics Today, 58*, 37–43.

Szilárd, L. (1961). *The voice of the dolphins*. New York: Stanford University Press.

Teilhard de Chardin (1999). *The Human Phenomenon (1955, in French)*. Eastbourne: Sussex Academic Press.

Tetrode, H. (1912). The chemical constant of gases and the elementary quantum of action. (in German) *Annalen der Physik*, *38*, 434–442.

van Hooff, J. A. R. A. M. (1972). A comparative approach to the phylogeny of laughter and smile. In: R. A. Hinde, (Ed.), *Non-verbal communication* (pp. 203–238). Cambridge: Cambridge University Press.

von Weizsäcker, C. F. (1952). *The World View of Physics*. London: Routledge.

von Neumann, J. (1932). *Mathematical foundations of quantum mechanics* (First Engl. edn. New York: Dover, 1943, Princeton: Princeton University Press, 1955).

Vrobel, S. (2011). *Fractal Time*. Singapore: World Scientific.

Waddington, C. H. (1941). Evolution of developmental systems. *Nature, 147*, 108–110.

Wallace, A. R. (1870). *Contributions to the theory of natural selection* (2nd ed.). London: Macmillan.

Wallace, D. (2012). *The emergent multiverse: Quantum theory according to the Everett interpretation* (1st ed.). Oxford: Oxford University Press.

Waterston, J. J. (1892). On the physics of media that are composed of free and perfectly elastic molecules in a state of motion, 1845. *Philosophical Transactions of the Royal Society of London A, 183*, 1–79.

Weyl, H. (1949, 2009). *Philosophy of Mathematics and Natural Science*. Princeton: Princeton University Press (first German edition: Oldenbourg Verlag, Munich 1928).

Weyl, H. (1952). *Symmetry*. Princeton: Princeton University Press.

Winfree, A. T. (1967). Biological rhythms and the behavior of populations of coupled oscillators. *Journal of Theoretical Biology, 16*, 15–42.

Winfree, A. T. (1972). Spiral waves of chemical activity. *Science, 175*, 634–636.

Yamaguti, M., Hata, M., & Kigami, J. (1993). *Mathematics of fractals*. Providence: American Mathematical Society.

Yandell, B. H. (2002). *The Honors Class: Hilbert's Problems and their Solvers*, pp. 159–162. Peters, Natick, Mass.

Zadeh, L. A. (1965). Fuzzy sets. *Information and Control, 8*, 338–353.

Zadeh, L. A. (1996). Fuzzy logic = computing with words. *IEEE Transactions on Fuzzy Systems, 4*, 103–111.

Zadeh, L. A. (2012). *Computing with words—principal concepts and ideas*. Berlin: Springer.

Zeilinger, A., Oxford open problems. See Briggs et al., 2013.

Zhabotinsky, A. M. (1964) A periodic oxidation reaction in the fluid phase (in German). In: L. Kuhnert & U. Niedersen (Eds.), *Selbstorganisation chemischer Strukturen* (pp. 83–89). Frankfurt: Harri Klein.

Zurek, W. H. (1985). Cosmological experiments in superfluid Helium? *Nature, 317*, 505–508.

Zurek, W. H. (2009). Quantum Darwinism. *Nature Physics, 5*, 181–188.

Zwicky, F. (1929). On the red shift of spectral lines through interstellar space. *Proceedings of the National Academy of Sciences, 15*, 773–779.

Zwicky, F. (1939). On the formation of clusters of nebulae and the cosmological time scale. *Proceedings of the National Academy of Sciences, 25*, 604–609.

Index

A. Sanayei and O. E. Rössler, *Chaotic Harmony*,
Emergence, Complexity and Computation 11, DOI: 10.1007/978-3-319-06781-0,

MIX
Papier aus verantwortungsvollen Quellen
Paper from responsible sources
FSC® C105338

If you have any concerns about our products,
you can contact us on
ProductSafety@springernature.com

In case Publisher is established outside the EU,
the EU authorized representative is:
Springer Nature Customer Service Center GmbH
Europaplatz 3, 69115 Heidelberg, Germany

Printed by Libri Plureos GmbH
in Hamburg, Germany